First sanctuaries, Jerusalem and Black Holes

References

1-E.Cassirer-Simbolo,Mito e Cultura-Ed.Laterza-Roma (Italy) 1985.

2-C.Skalicky – Alle Prese con il Sacro-Herder -Roma 1982 .

3- F.Barbiero –Alla Ricerca dell'Arca dell'Alleanza – SugarCo-Milano(Italy) 1985.

4-D.G.Bressan – Samuele -Marietti-Torino(Italy) 1960.

5-A.Mistrogiro-L'arte Sacra-Messaggero-Padova1983.

6- G.Damasceno – Difesa delle Immagini Sacre-Città Nuova Ed.-Roma 1985.

7-A.Lancellotti et al.- Collectio Assisiensis,8,Studio Te ologico"Porziuncola"-Assisi(Italy)1971.

8-M.Grant-L'Antica Civiltà d'Israele-Bompiani-Milano (Italy)1984.

9-F.Brown et al.- Hebrew and English Lexicon of The Old Testament-Oxford University Press-London 1962.

10-C.M.Martini-cardinal-Chiesa-Vescovo-Martirio-Paoline-Roma 1983.

11-A.Issar- Fossil Waters Under Sinai Peninsula;see Scientific American July 1985,p.82.

12-Encyclop.Amer., 21- American Corporation - New York 1977.

13-J.Perrot-Siria-Palestina,1,Nagel-Ginevra- 1977.

14-E.Picutti - Storia del Triangolo Numerico; see "Le Scienze"Gennaio 1984-Milano(Scientifc American).

15-E.Fiandra- <Le Scienze>,Aprile 1983 –Milano-Italy-(translated Scientific American).

16-J.Oates-Babilonia-Newton Compton Ed.-Roma .

17-A.Parrot-Archeologia della Bibbia-Newton Compton -Roma 1978.

18-A.Kempinski-Siria-Palestina,2,Nagel-Ginevra 1977.

19-P.Maffei- La Cometa di Halley - EST Mondadori - Milano 1984.

20-A.Cotterel-Enciclopedia delle Civiltà Antiche-Editori Riuniti-Roma 1981.

21-E.R.Galbiati et al.-Atlante Storico della Bibbia e del l'Antico Oriente-Jaca Book-Roma 1983.

22-R.Harker-il Mondo della Bibbia -Newton Compton-Roma-1983.

23-N.Borrelli -Tradizioni Aurunche-Centro Studi Min -turnae-Perugia 1984 .

24-B.Hutchinson - Orologi Antichi -Mondadori-Milano 1982;

A.Paul - Photo-Guide de l'Ancien Testament - Fleurus-Paulton 1976.

25-K.Crime et al.- The Interpreter's Dic. Of The Bible-Supplementary Vol.- Abingdon Press- Nashville 1976.

26- The Interpreter's Dic.of the Bible- New York 1962.

27-Supplements to Vetus Testamentum-J.Brill-Leiden 1960.

28-J.B.Meetzlershe-Real Encyclop.-1927.

29-M.Haran - Tempels and Temple Service in Ancient Israel-Clarendon Press-Oxford 1978.

30-J.Mauchline - 1 and 2 Samuel - Oliphants-London

31-R.W.Klein-1 Samuel-World Books Publisher-Waco (Texas)1983.

32-J.A.Emerton-Studies in the Historical Books of the Old Testament-J.Brill-Leiden 1979.

33-R.E.Brown- Comentario Biblico San Geronimo ,1, Edic. Christiandad-Madrid 1971.

34-R.K.Harrison-Introd. To the Old Testament-Tyndall Press 1970.

35-H.M.Barstad-The Religion Polemics of Amos-J.Brill - Leiden 1984.

36-G.Von Rad-Teologia dell'Antico Testamento-Paideia Ed.-Brescia(Italy)1972.

37-E.Pace-Dizionario di Sociologia e Antropologia Culturale-Cittadella Ed.-Assisi(Italy)1984.

38-J.Perrot-Siria-Palestina,1,Nagel-Ginevra Swissland 1977.

39-V.Malka-Israele-Edizioni Futuro-Verona 1982.   -

40-E.Beaucamp-La Bible et le Sens Religieux de l'Univers-Du Cerf Ed.-Parigi 1959.

41-R.De Vaux- Ancient Israel-Darton Logman & Todd-London 1962.

42-P.Reymond -L'Eaux,Sa Vie et Sa Signification Dans L'Ancien Testament-J.Brill-Leiden 1951.

43J.Uris et al.- Gerusalem-La Cantique des Cantiques-Doubleday-New York 1981.

44-T.Ballarini- Salmi – Dehoniane – Naples(Italy)1978.

45-A.Dupont - The Essen Writings from Kumran-The World Publishing Comp.-Cleveland 1962.

46-A.Gelin- Les Idées Maitresses de l'Ancien Testa - ment -Du Cerf-Parigi 1959.

47-J.Dheilly- Dictionnaire Biblique - Desclée - Tournai 1964.

48-M.Gell'Mann et al.- La Natura dell'Universo Fisico- Boringhieri-Torino(Italy) 1981.

49-M.Gardner- L'Universo Ambidestro-Zanichelli Ed.- Bologna(Italy)1984.

50-Suppl. a <Spendibene> anno XV-il Nuovo Commer cio n.16 del 31Marzo 1990.

51-B.Cester-Orologio; Scienza e Vita -5 maggio 1990.

52-A.Caproglio et al.- La Natura, l'Uomo, la Scienza - vol.3 -Editore Morano-Naples(Italy) 1980.

53-A.Laepple- il Messaggio Biblico per il Nostro Tempo -Paoline-Roma(Italy)1984.

54-D.Rops-La Vita Quotidiana in Palestina al Tempo di Gesù-Mondadori –Milano(Italy)1986.

55-G.Herm – I Bizantini-Ed.Garzanti-1985.

56-G.Romano- Introduzione all' Astronomia – Franco Muzzio Editore-Padova(Italy)1985.

57-G.Hegel- Le Orbite dei Pianeti-Laterza-Roma 1984.

58-Eusebio-La Préparation évangélique-Du Cerf-Parigi 1975.

59-F.Pierini-Le Religioni dell'Antichità; see Guida alle Religioni-Paoline Ed.-Roma(Italy) 1985.

60-T.Reinach- Antiquités Judaiques – Leroux – Parigi 1912.

61-J.Gutman - The Temple of Solomon -Scolar Press-Missoula 1976.

62-D.Gottlieb- il Giudaismo ;see Guida alle Religioni-Paoline Ed.-Roma(Italy)1985.

63-A.Cotterel – Civiltà Antiche – Editori Riuniti-Roma 1961.

64-A.Parrot-Le Temple de Jerusalem-Nestlé Delacaux -Neuchatel 1956.

65-P.Vanderberg-La Maledizione dei Faraoni-SugarCo -Milano(Italy)1985.

66-A.David -L'Egitto dei Faraoni-Newton Compton Ed. - 1975.

67-F.Nera -Guida alla Civiltà dell'Antico Egitto Monda dori Ed. –Milano(Italy)1985.

68-Dizionario Biblico-Diretto da F.Spadafora-Editrice Studium-Terza Edizione-Roma 1963,p.587 .

69-Bernhard W.Anderson-Understanding the Old Testament - Fourth Edition - PRENTICE HALL - Englewood Cliffs-New Jersey-07632-

Introduction

In its first part, this work is a description of Jerusalem's temple,of its proportions and of some more an
cient shrines as well laying on the promised land.   -

   After the event of the Jewish settlement in land of
Canaan, we see stable structures at Shiloh and else -
where.To take a survey of the numerical data concer
ning the sacred area is necessary,because their criti -
cal examination has never been done. Moreover a -
drawing is endowed with a significance that one can't
completely grasp (p.54;p.88).   -
Surveying is also useful to compare thecnics of diffe -
rent epoches.Sometimes,through the connections in
a plan, one could follow steps of a progress. We find
Shiloh,ark's first venue and soon after pagan temples.
The use of numerical expressions is generally modera
te,because one doesn't want an overall view with ma
ny details.   -
  The architect has to think how to put an ordered ar-
rangement of a plan,suitable for a whorship place.In
the chapter involving non Jewish temples, columns' -
groups cast light beams on the earth.These originate
from different celestial sources. One can infer this -
phenomenon[1] at Biblos (p.37) and Hazor(Kempiski Sy-

ria-Palestina 2-Nagel-Ginevra 1977).Egypt's pyramids were also used for the same purpose(p.107).   -

In the second part,after the patriarchal time,when the first altars were built,it's brought forward the pro gressive diffusion of Jahweh's cult through pagan are as already sacred for other peoples.Temple's theolo - gy is explained with the concepts of covenant, cho - sen people and priesthood,without overlooking the - view points of both ecclesiology and eschatology.   - Space is given to a cherubim's description, taking into account what is known with the trinitarian theology - from s.Basilius'time(4th century).   -

Some of the more prominent novelties come in this treatment are the following: - 1) A verification that Moses' tent is associable to a - summer solstice event(p.52);one really singles the an gle through $47° = 2(23°.5)$out.   - 2) Aijalon and Gibeon lay on the same parallel.Then a proof is made that <sun stand still at Gibeon, and - moon, in the Valley of Aijalon (Jos10.12) indicates - noon ;it's a correct statement to describe sun's appa rent motion. In fact at the end of its ascent to noon, the sun makes for the Earth because of the sunset.   - 3) We show a reconstruction of Dionigi's ancient ca lendar useful also today .   -

All misunderstandings about the figure 753 as beginning of its dates fall down[see the newspaper "il MAT
TINO"-Naples(Italy)Dec.10-2000].   -
4) With temple's destruction the city gained in presti
ge;it's important the gathering on the mount(p.54). -
5)One gets a deciphering of an egyptian numerical ta
ble,by explaining its astronomical content(p.110).   -
6) It's shown that Ezekiel's chapter n. 20 conveies a -
coded message concerning calendars; arguments at -
(p.111;112).Analogous dates'cheks should be extend-
ed to Gospel(p.20).   -
7) There is a system of equations revealing the distan
ces of some galaxies from a singular point,and also of
stars as respect to sources of attraction (p.114).

   My warmest thanks to both Prof.Pasquale Colella of
Laterano's papal University(Rome)for his invitation to
study the Jewish first temples and Prof.Gaetano Sàvo
ca of s.Luigi papal University (Posillipo-Naples) for ex
planations about the structuralist biblical analysis.   -
Vittorio Italo Morrone
Castel di Sasso- 81040 -Caserta(Italy),Gennuary 2012.

Art's language

As for the convinctions' change about the aesthetics, Cassirer puts it:<the language world must be studied as an instrument of the human way of thinking which helps us when an objective world is under construction[1]>. This experience is regarded as a phenomenon - where the subject who observes an art work doesn't passively remain recording images of the surrounding world ,but takes part in the motion fixed by rhytms,li nes and shapes that constitute the beauty according to the classic conception,in their unity.   -

Dealing with a subject concerning temples one will see that the attempt to represent the sacred has been carried out. It's a feature of the artistic work to suggest the content of emotions, of events one lived through.In an Eliade's work about the morphology of a sacred space it's also symbol,because it reveals a sa cred reality;therefore implies religion's memories.   - So, all what you find in it paves the way for a speech going from the finite towards the incommensurable - dimension of the religious spirit. If in myths one finds values concerning vicissitudes occurring in the world, an out of use by now statue supplies information a - bout the ancient times and human nature.Then,to - put a separation between mythical reality and the sa-

cred one has to remember the cosmogony; in fact, e -
very mythology can be traced back to it .   -
          Now we analyse the ideas spread on the land of
Canaan and neighbouring countries from 1234 B.C. to
1183 B.C.;the first date corresponds to Israelites' certain presence on the land of Canaan,while the other ,
concerning the accession in Egypt of Ramses 3°,marks
Jordan's crossing and nomadism's end[3]. Introducing
Shiloh won't simply mean to give a description of the
territory.   -
Samuel's books tell us the main events of the monar-
chy's institution.David represents the most important
element.His talents,meekness and magnanimity,mag-
nificence of his reign and knowledge of what he had
to forgive will almost make him a Messiah's prefigura
tion[4]. Thus one has a reference to the transcendent,a
first indication of the Church,<the humanity raised to
the supernatural plane>[5]. In other words,the temple -
becomes symbol of a salvation's plan.Saint Fathers -
have long spoken about the incisiveness and role of -
the sacred art. Areopagite s.Dionigi[6] makes reference
to the symbolic representations that < stem from the
divine characters and are copies and manifest images
of the supernatural arcane visions>.In the temple the
re is copy of divine advice and one must specify that
every sacred image is in its turn image's copy ,becau -

se the divine will,expressed through commandments allowes to have only a shadow of the future goods.– This is a concept widely developed by Damascene s. John[7]. Actually with the temple one has more, becau se there is Jahweh's presence in it.According to the rabbinical tradition[8],the manifested glory of God (so - metimes called his shekinah glory) descended on to - the tabernacle. The Lord listen to the supplications - and intervenes in the history .The term shekinah was coined when the building had been destroied;derives from sakan(to dwell);indicates that God listens to his people. Paradoxically called Lord of hosts(or of hea - ven's armies) (Am 4.13), at the same time he is God - of peoples and God of the virtues;but essentially the Almighty God who gives peace. With the sanctuary, - historic events are remembered.The sense of great - ness sent from its sensible sight forth is a precursory indication of the free gift revealing God's saving bene volence. It always come in useful exciting emulation - while the works done by the Creator are related.   - Therefore it won't be a surprice if on the New Year's Eve party,when lunch is over,the head of a Jewish fa - mily distributes a pomegranate fruit to all.   - There is a prayer that as many good actions be made during the year as could be suggested by the inner - structure of this fruit. It decorates the capitals on the

temple front and is part of the cosmos which reveals
God's goodness.In the Canaan land El Elyon is the su -
preme being. To him,different names are attributed ,
such as El-Olam(=ancient of ancients)[9] and El Bethel.
   Notwithstanding this,a polytheism exists for which
the codes'ethics doesn't reflect a devotion's sincere
feeling.Therefore there is no precepts' provenance or
their foundation on a spiritual cult which is renewal -
of the heart .The creation's concept separates the re
ligion from false speculations.   -

Matrix for salvation

From the prophet Ezekiel one gets the following prospectus of the passage from a promise to its accomplishment.

PROMISE(Ez.20.6):   **Existential sector**(Ez.20.11)-Law addressing   1)the universal community(Ez.20.19) <Walk taking into account my statutes, observe my - laws and bring them to effect>.   2)Starting from this singular intervention,one keeps on with teaching in - order that the individual live in it through:a)(Ez20.39) removal of misdeeds(redemption);   b)Holy spirit's gift[<you won't profane my Holy Name>(salvation)].
Cosmic sector(Ez.20.12)-   The Saturday,and this r<u>e</u> - veals itself: (Ez.20.40)at the mountain(SIGN); (Ez.20.41)with gathering(WONDER).

ACCOMPLISHMENT AS GOD'S WORK

w_1)**Temporal**(=priest's ministry)   Ez.20.7:Throw the idols away].WAY(wonder):<I made them go out of Egypt>.   RESULT(accredited):<Then you will know>.
HUMILIATION(Ez.20.22):   <The sons rebelled against me,but I held my hand back>.

w_2)**Definitive**(=eschatologic)   A)Triumph(Ez.20.35)- (Ez 20.42):<I judged to intervene>.B)Lordship : WAY (Ez.30.37);yoke(=guide).RESULT(Ez.20.41):Holy in the peoples' opinion.

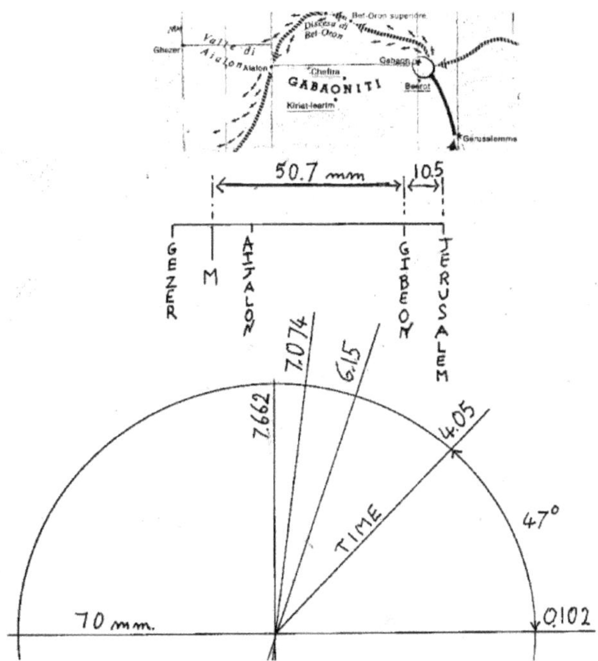

Scale: After 360° one adds 30.24 along the circle.

Fig.1 –Above:Relative longitude[10] concerning Gezer, Gibeon and Jerusalem.
 In Ez.20 one notes two lexical fields(time, place) and four roles (Name,cult, life, appeal).

$$(7.662-4.05)=3.612 \Rightarrow \frac{30.24}{3.612}=\frac{360°}{43°}$$

$$19.002-4.05=14.952 \; , \frac{30.24}{14.952}=\frac{360°}{178°}$$

17

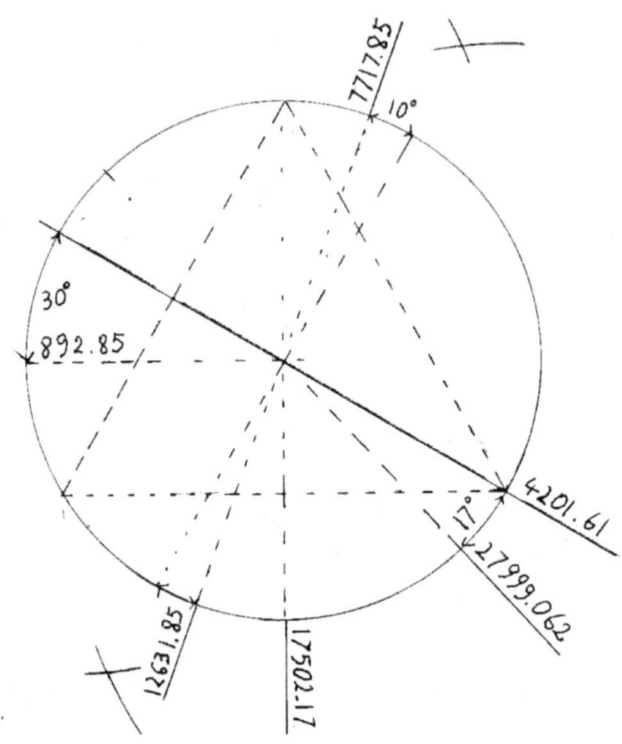

Scale: 360°≡30.24;

Fig.2a- Periodic data on Mount Zion(p.111).

$$\frac{2.73}{2.73(35)(161}=\frac{32°.5}{183137°.5};(23101.14 - 7717.85)=$$

2.73(35)(161)

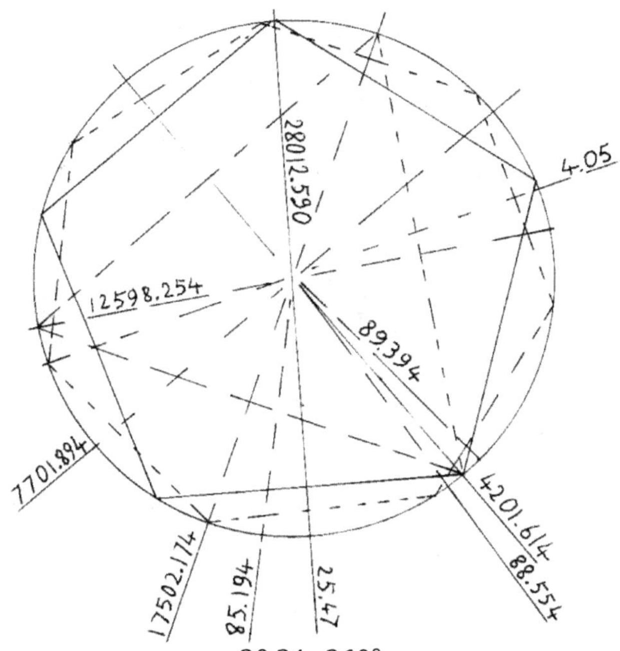

Scale: $360° \equiv 30.24 \Rightarrow \dfrac{30.24}{0.504} = \dfrac{360°}{6°}$

Fig.2b-Starting point **4201.614** (p.112).

$92.01-88.554=3.456=3.024+0.432 \quad \dfrac{30.24}{7(0.432)} = \dfrac{360°}{36°}$

$70(\mathbf{365.2422}+92.01)=32007.654$

$\dfrac{30.24}{32007.654} = \dfrac{360°}{381043°.5}(85.194-0.504)=84.69$ (p.117);

$(\mathbf{28012.59}-84.69)=\mathbf{27.3}(1023)$

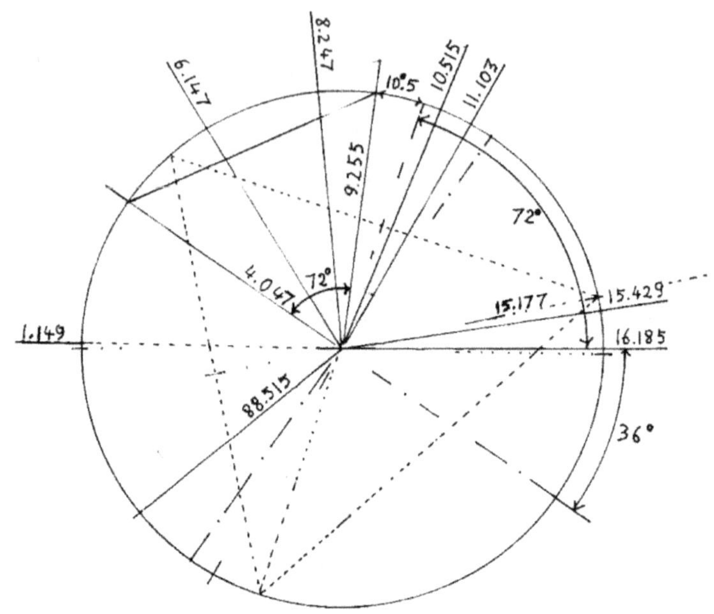

Scale: 360°≡30.24

Fig.3a-Relevant points in Mark's Gospel *
     [(4.047-9.2559)=linkage Ez.-Matt.]

$(15.177-4.047)=11.13 \Rightarrow \dfrac{30.24}{11.13}=\dfrac{360°}{132°.5}$

Mk 1.14-After John had been put in prison, Jesus - went to Galilee and preached the Good News from God.

Mk 16.18- If they pick up snakes or drink any poison, they will not be armed, they will place their hands on sick people,and they will get well.

*Benoît Standaert- il Vangelo secondo Marco - Borla Ed.-Rome 1983,p.22 .

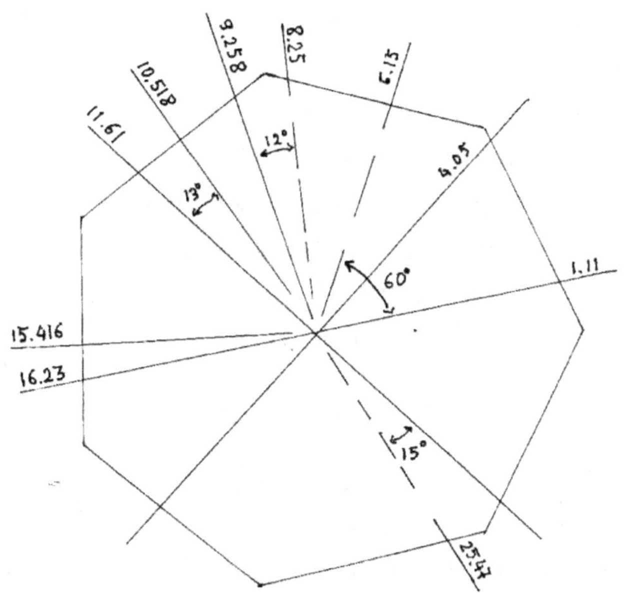

Scale: For an angle through 12° we put a quantum change of 1.008 along the circle;

$$\frac{30.24}{1.092}=\frac{360°}{13°}\quad,\quad \mathbf{2548}=7(364)\ (p.54)\ .$$

Fig.3b-Reference points* in Mark's Gospel .

$$\frac{30.24}{9.258-1.11}=\frac{360°}{97°},\ \frac{30.24}{25.47-10.518}=\frac{360°}{178°},\ \frac{30.24}{89.394-25.47}=\frac{360°}{761°}$$

$$[(11.1\text{-}9.13)=1.97\Rightarrow \frac{30,24}{1.97}=\frac{360°}{23°.45238095}\ ]$$

Mk.1.11=Voice from the sky at Jesus' baptism.

*Benoît Standaert,p.22.

Sun,stand still at Gibeon ,and moon,in the Valley of Aijalon.

As Gibeon and Aijalon lay on the same parallel[10] it's easy to imagine an angular distance (= longitude) - between these two places.In accordance with Hoyle (physicist),one puts 56 holes[11] along a circle.

From his data,the shift through two holes a day of moon's motion is $\frac{2}{56}(360°)$ or $\frac{360°}{28day}=\frac{12°.8574236}{1day}$.

$\frac{360°}{364day}=\frac{0°.98901989}{1day}$ is the sun's change in a day,arch - quoted in the Bible as < (angular) space for a whole day>. Let's assume (Fig.1,p.17) on the parallel passing through Gibeon,as point chosen in the Aijalon Valley, the middle point M of the distance between the meri dians of Gezer and Aijalon.To find GM(Fig.1)we shall - use the formula with $α=\frac{l}{r}\cong\frac{GM}{r}$ where

r=56(21.84)=1223.04 Km ,

and $α\cong0.01726$ radiant,owing[12] to the proportion*

$$\frac{360°}{0°.989}=\frac{2\pi}{0.017261306}. \qquad [* \ \pi=3.1415927]$$

GM$\cong$0.01726(1223.04)$\cong$21.109Km

The biblical atlas'scale[10] suggests that 12mm. are equivalent to 5Km;hence

$$\frac{5Km}{21.1Km}=\frac{12mm.}{50.64mm.} \qquad [\textbf{2184}=6(\textbf{364})=42(52)]$$

The direct measure is 50.7mm on the atlas.

Modern method:

For the sun, $\dfrac{360°}{365.2422\,day}=\dfrac{0°.985647332}{1\,day}$,

$\dfrac{360°}{0°.985647332}=\dfrac{2\pi}{0.017202791}$,and the formula $\alpha=\dfrac{l}{r}$

becomes $0.017202791\cong\dfrac{GM}{1224.596}$;whence

GM=21.06646905Km

because $r=1224.596=\dfrac{6367.8992}{5.2}$.

From Earth's diameter at the equator and the other measured from pole to pole ,

$\dfrac{12.757+12.714}{4}\cong6367.9$Km =**radius of our planet***

(*W.Schroeder-Astronomia Pratica-Longanesi Ed.-Milano 1982,p.14).

Finally $\dfrac{5Km}{21.06646905Km}\cong\dfrac{12mm.}{50.5595mm}$;

(50.7-50.5595)=0.1405mm,

$$\dfrac{12mm}{0.1405mm}\cong\dfrac{5Km}{0.0584Km}$$

We recall the main points of (Jos 10.12):
Sun stand still at Gibeon[13],and moon (remain as fixed point) in the valley of Aijalon.....and the sun stood still in midhaeven (at Gibeon).<(Then it)did not hurry to - set for about a whole day> means that moon and sun are mutually aligned on the Aijalon Valley **after** the sun's motion through 0°.989(= a whole day).   -

Thus sun's motion from the zenith at Gibeon through 0°.989(towards west) transfers the sun from Gibeon - to the valley(point M).Afterwards the sunset will take place after six hours.   -

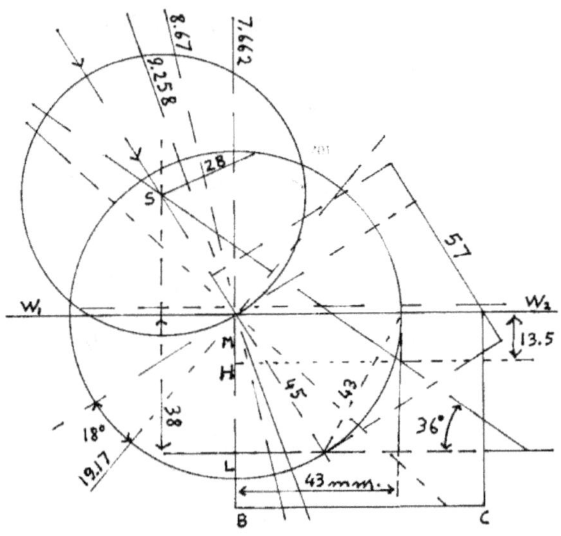

45mm|146.25
43mm|134.75
Fig.4-Measures[14] at the Gibeon pool[*].
ML=pool's wall;BC=a subsoil level.
W_1W_2 =Earth's surface .

$$\frac{30.24}{134.75-89.39}=\frac{360°}{540°} \qquad \frac{30.24}{7(146.25-7.662)}=\frac{360°}{11549°}$$

$(\mathbf{16.44}\text{-}8.67)=7.77 \quad \Rightarrow \quad \frac{30.24}{7.77}=\frac{360°}{92°.5}$ (p.117) .

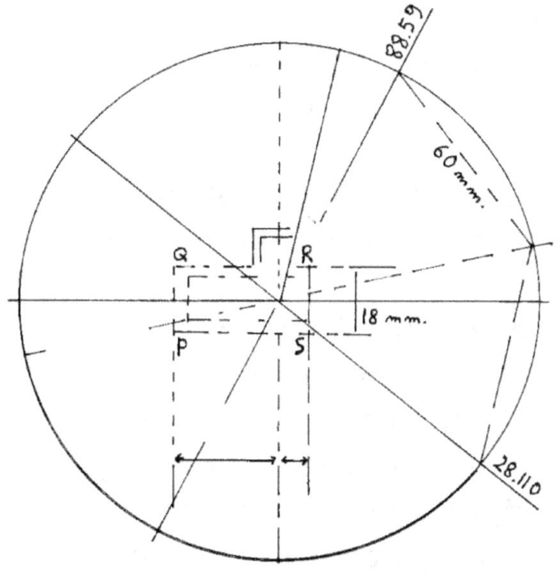

Scale:211.68 is equivalent to 360°.

⇒By rotation through 12° one adds 7.056 along the circle.

$$\frac{211.68}{88.59-28,11}=\frac{360°}{2(\frac{360°}{7})}$$

Fig.5- PQRS, proportional enlargement of the inner - rectangle,is   Ezekiel's theoretical sanctuary.Here one confronts it with s.John basilica's plan [15] (Rome).   - 88.59 = 3(29.53); at multiples of 29.53 days,sun and moon are aligned(p.100;p.97).   -

Circle's radius=70mm; 60mm.=eptagon's side .

A stop at Shiloh

At present,in Shiloh's district, at 680m,one finds no - thing but a sense of rough desolation,sometimes bro- ken by the flora's splendid enchantment of not far- off resorts.   -

 One is on the promised land that flows with milk and honey. At 209 m.of height,ideal for meditations[16], it is still rich in vineyards.   -

Here the priesthood expresses itself through one of its fundamental components,in which the prophet is seen as the man to be consulted for an oracle.The - re is in vogue the practice to obtain divine communi - cations called Urim(the lots stored in the priestly e - phod). It doesn't substantially reveal more than the - well established mediation function. Zuph, Elkanah's district,means honeycomb[17] and one infers that the loss suffered with ark's capture is symptom of a spiri- tual decline in which one doesn't have accordance - with what was decided upon from a supernatural ar - cane plan.   -

The clash with sea peoples' armies stresses the importance to find a way to get peace with a mutual respect. Beth Shemesh will be set as border between two opposing stoks, and Shiloh leaps out at you with its peaceful sacrifices open to brotherly banquets.   - Jerusalem's temple introduces you to breathe the me

mory of Abraham, whose devotion is admired and re
minds the various offerings[18].   -

The environment trasmits achievements' heritage
in which intellegible things gain visible elements: defi
nite lines suggest what is immaterial. The construc -
tion will be completed on that Easter after which Paul
will be able to speak about the new stones put by -
him;of the blood poured as libation on the sacrifice
and offering of those who will believe in the risen –
Christ ,who is the corner stone. Zuph recalls Moses' -
memory,his life,the bitter waters (with sulphates) of
the sources that after him were called[19] and the tuffs
of neighbouring places.With augers have appeared -
liquid's fossil layers laying under the Sinai subsoil -
with a silent drainage up to the Dead Sea shores,not
far from the 32nd parallel,where lived Samuel and a
man called Elkanah, actually gained by God[20].   -

The environment
In a recent census (1983), Jewish State's population -
was about 4.000.000.The territory's evaluation yields
21.000 Km^2 as area.   -
It's a transition strip between mediterranean and arid
climate[21]: one passes from coast's nearly tropical as -
pect to a continental with temperature's considera -

ble changes between day and night[22]. In the valley of the Jordan one has high temperatures also in the sha dow. On the mountains, at east and west,rainfalls - such as 600mm. per year are found, reduced of a fac tor ten elsewhere: one has damp heat and it hardly rains during the year.Characteristic is the Dead Sea - plain[23].Its coasts,nowadays occupied by thriving indus tries making use of its minerals and salt characteris - tics, pave the way for technology .   -
The landscape had to appare different enough to an ancient inhabitant ,nomads and invaders.All the surfa ce undergoes geological changes also registered from hills' tops with planes at distinct levels.   -

   With archeological excavations one gets an outlined scene of the installed populations. For Palestinians in the south one has a documentation since 3.000 B.C.; then semites have been reported in Egypt, while Ba - bylonians rule in Canaan.   -

 As for Syria,some groups are formed by Arameans in 2.000 B.C.; the stay in the palestinian territory of the- se,of canaanites and other groups doesn't give rise - to arguments while Abraham ,under the pressure of a famine,goes into Egypt. This is indication about the peaceful character of inhabitants.   -
Yet, hard fights won't be lacking among local tribes.- Now we introduce the palestinian habitat with some

images.For the conquest time we recall Gilgal and Bet Oron,a few khilometers from Jerusalem.

There are high areas where take root olive trees and cypresses.In the extreme north,mount Hermon with a changeble climate when one proceeds from its ba se toward the imposing top.At south,one finds Masa da,cliffs and desolate ravines;and then,going towards north, En Gedi on the western banks[21] of the Dead - Sea. On year's first months the land is covered with - flowers. At Arad one has an expance of sand and gra- vel.Then precipices are admired.

An almost torrid heat is stifling after April. Part of the fauna is described in the Song of Songs. With Masa - da's fortress we think back many Jewish expressions - referring to Jahweh. Samuel's mother, Hannah, will - speak of God as a cliff: He is Israel's rock.

It's quite easy to think of a people devoted to sheep farming and we have direct evidence of this activity - for 4.000B.C.;nevertheless the shepards' nomadic cha racter is more manifest at Beersheba.   -

Moving from a place to another becomes necessary owing to climate's features: 200mm. of rainfalls is - what the agriculture can barely endure[24].   -

With the presence of goats, pigs, oxen we find mel ting pots ,fireplaces and tools to perform basalt's cut ting. At En Gedi, wheat and barley abound. Of the -

country,also known for gazelles , will speak Ezekiel to point out which power is implied in God's blessing, - owing to temple's existence at Jerusalem :fishes   will be abundant even close to the Dead Sea.   -
 In Egypt and Palestine there were cultivations preced ing the 4th millennium.   -

   As for the habitations obtained from rock,there are configurations with definite shape. Evidences for the study of the rectangular triangle are given in Fahun's papyrus for the beginning of the 20[th]century B.C.; py- ramids belong to a more ancient time[25].   -

Observations

One wants to survey the temple's dimensions (see p.49,p.53,p.118) . The environment and culture for that time are also implied.A trace can be found in the structure of buildings.As it will be clear later,one dis- covers elements of a same mosaic in various places.

          In practice,the culture spread on a wide area becomes only different for original local features. As - we mentioned the Jewish people, it was also necessa- ry to stress divergent points of view emerging from - the cosmogony.   -

In the index of this work one immediately gets a new
ness' impression for singling measurements'elements
out concerning both space and time.To have an idea
about times' evaluations one can soon think to the ca
lendar.A first presentation of Ezekiel's temple in relation with cardinal points reveals a linkage with an angle through 23°.5 which marks a season's start in astronomy. One recovers the meaning of<sun stand still
at Gibeon, and moon, in the Valley of Aijalon>: it is a -
(kinematic)right description for zenith's direct obser-
vation: at noon ,in both Gibeon and valley comes to -
an end sun's apparent run. Aijalon and Gibeon lay on
the same parallel enabling us to locate Aijalon Valley
as the middle point (p.17) between the meridians of
Gezer and Aijalon.

    To study the noon hour one needs a sundial[26] in
which use is made of the latitude of the place where
it must be put;29.53days= $\dfrac{88.59}{3}$ bring about[27] an align
ment of sun and moon (p.100;p.97).   —

Canaanite shrines.
Parallely to the buildings growth one observes a thec-
niques' change through the local industry's products
such as bronzes ,ceramics and what can be obtained
from written textes come to light.   -
One can't deny sign's value of the architecture,becau
se it is a synthesis of a parts' series whence the social
situations'aspect comes out.   -
For Palestine,between 1234 and 1183B.C.,we need -
to remember problems concerning the pagan tem -
ples.We must turn our attention to the religious conceptions of Mesopotamian peoples,particularly of Su-
merians,whose licterature was known in Babylon,and
of Chaldees[28]. Both of them lived near the estuaries -
of Euphrates and Tigri.   -
   Babylonian poems have been found at Niniveh ,the
Assyrian capital with Assurbanipal's famous library -
discovered in 1853. The gods,according to Chaldees
can intervene both in favor and against men. -
 What they thought about the afterlife is contained in
the Epic of Gilgamesh,king of the ancient city of Uruk.
   One made already use of asphalt also available on
banks of the Dead Sea,and in the poem are described
date's vine,metals such as gold and silver,sacrified o
xen and sheep[29]. The god at the time of creation is -
Marduk ;men, made into a mixture with this god's -

blood,must practice the whorship in a place where -
the heart rejoices at prayers.   -
     There was magic. Soothsayers and astrologers did
not lack.In the Babylonian written documents there a
re normative indications for food,dresses,instructions
and prayers. The spiritual phenomenon's perception -
is known; nevertheless speculations are present as it
happens in other religions too.   -
The objectivity of the transcendent dimension which
is integral part of the human being leaks out of the
provisions,and one finds an existence's subordinati_-
on to the constitutive laws inscribed in the cosmos. -
Moreover one mustn't forget Babylonians' mithology
exists next to other myths.Phoenicians' theories must
be kept an eye because there was a strict linkage be_-
tween them and Palestinians.In ancient times one al_-
ready thought a demonstration that Greeks' pan -
theon could be derived with a simple tranformation -
of historical events handed on from Phoenicia.   -
Eusebius gives news about snakes'cult practiced the_ -
re,of allegories and initiation rites for mysteries. A typical decription of mystagogic praxis can be found in
Plutarcus (first century). No real research of truth is
found;it's essential testing instead of learning.   -
     Mari,one of the greatest cities of the world, was
destroied in 1759B.C. by Hammurabi's armies. Then

Hazor was the most important palestinian city. It has been located near to Tell el-Qedah according to indications of Yadin(1955);close to lake Huleh. One of its kings, Yabin,will be defeated by Joshua.

 Hazor provides a precise documentation for the typical Canaanite city. Also in Biblos there are constructi ons dating about 1786 B.C.; in Egypt the 12th dynasty is ruling.There was a blooming trade.

In the bronze age, the temple is equipped with cell, - and long since the city can be considered the most flourishing both of the Syrian dominion and Palestinian one.In its temples have been found egyptian pots, seals and objects referring to contacts with Mesopta mia.Colored ceramics of Hazor and Dan appear in the middle bronze age. We find also a variation of fortifi cation's kind.Hyksos live in Palestine during the early iron age. At Shiloh,there will be those constructions and typical ceramics spread all over the Galilee ;but in Ai and Negev the houses are often with four rooms and amphoras show a curved border.Trabeations are sustained by pillars and one builds on canaanite ruins at Shiloh[30].

Afterwards we find Hittites in Palestine and there will be an Egyptian rule. From documents regarding the - first people we get news about Philistines'beliefs; so -

we know the general attitudes about war's spoils and this is useful in the ark's restitution.   -
A comparison among correspondent gods is effective to understand propitiations or aspects of the sacred.- The practice to carry god's images of enemies as vic - tory's remains allows also us to perform a thelogical reflection.   -

In 1Samuel,where prophetism is stressed,his work will be showed up as a result of Yahweh's presence. - Aegean population's influence, especially by Micenae is accounted for with the examination of the finds dis covered in sites. As far as south one has characters'u niformity.   -

Other important resorts
At Sittim,east of the Jordan,Joshua waits for the spi - es' return before moving toward Gilgal[31].Jericho's - conquest is immediate.It was a small village occupy- ing an area whose greatest dimension reached[32] a bout 230m.On the main road there were houses and walls made   of bricks. Among the finds come out of - the ancient Tell(=remains)are enclosed millet, barley, lentils,grapes,cloths and various shape's vessels, po - megranate fruits,wooden combs, pendants, painted - masks used for copses and perfume's little bottles. -

Thus one has an outline of what was available in a -
country with bananas in december,date palms,owing
to sultry weather(with 35°C on summer).   -
It's proved there was wine's carriage to close resorts.
For this reason Hannah was considered drunk from -
Eli: during the religious festivities one made use of al
coholic drinks. Pressure was obtained under a flat sto
ne,rudimentary press.In Romanians' age it will be still
in use a soaking vat covered with wooden board pres
sed by feeth.This will be recalled,in the Virgilian Geor
gic poem[33]. On the Palestinian hills the product was -
kept in little containers put in rocky cavities.The whole had to be covered to avoid temperature's changes.
   A Gibeon's monument worth to be mentioned is a
pool[37](p.24),substantially a cylinder digged in the flat
soil. In case of siege the access was allowed through -
an underground tunnel. In general,columns or stones
are the means used in antiquity to transmit a memo-
ry.Their inscriptions relate noble achievements of her
oes, days of agreements,cult in a temple.

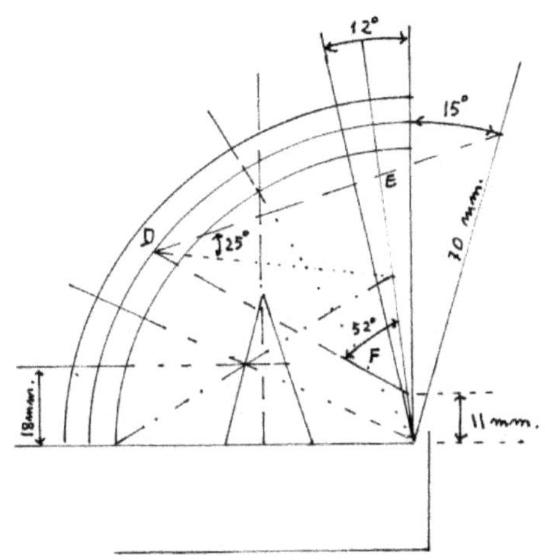

Fig.6-Measures[34] of space and time at Biblos.

DEF is L.Danzer's triangular tile*.

The distance between pyramid's vertex and axes' ori gin is 49 mm.

$(\mathbf{553}-49)=504 \Rightarrow \dfrac{30.24}{504}=\dfrac{360°}{6000°}$ (p.110)

49 |21.8

18.05 |8.03

36.1 |16,06   If $\tan\alpha = \dfrac{8.03}{16.06}$, $\alpha=26°.56505118$

$90°-12°=78° \Rightarrow 78°-\alpha=51°.43494882 \cong \dfrac{360°}{7} =$

  $=51°.42857143$

*M. Senechal- Quasicrystals and Geometry- Cambridge University Press-New York 1955 .

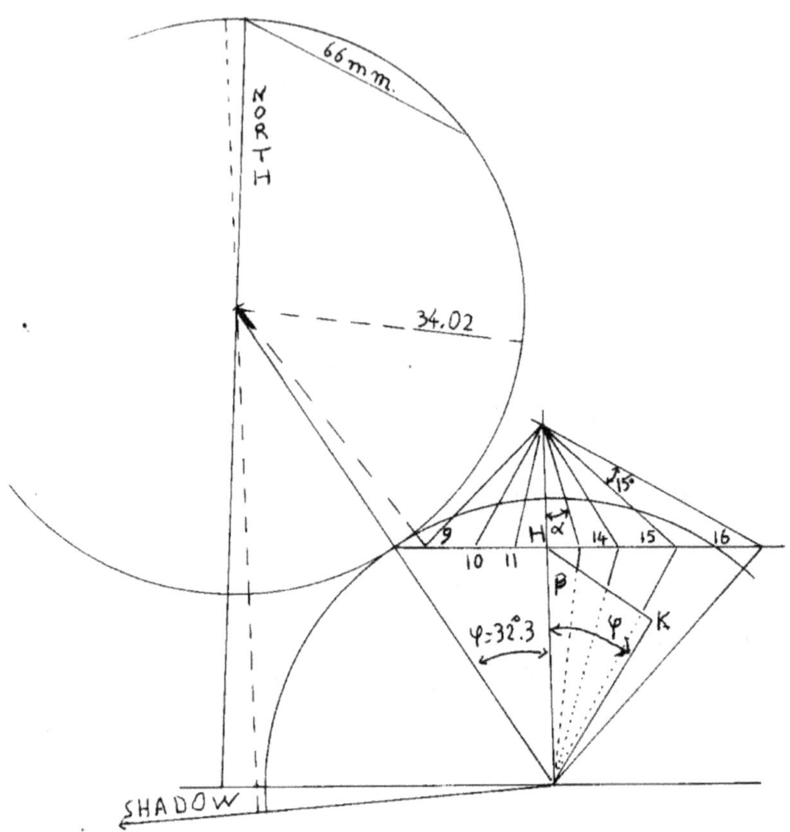

Scale :   12°≡ 1.008

Fig.7- A Gezer 's colonnade[35] casts a shade which al‌-
lows to locate the north pole. Here it's put in corres‌ -
pondence with a horizontal sudial drawn in the verti‌ -
cal plane);$\alpha=15°$.[φ=latitude;$\tan\beta= \sin\varphi\tan\alpha$ (p.105)].
66mm is eptagon's side for the circle whose radius is
77mm.   **3402**=(30.24)112.5 $;\dfrac{30.24}{34.02-7.56}=\dfrac{360°}{315°}$ (p.113).

Jewish first temples

During the wanderings in the desert ,the presence of Yahweh was associated with the moving tabernacle ; however Shiloh will be the par excellence temple be - cause there was the ark .Today, we only find ruins of the village,at 18 and a half miles to the north of Jeru- salem on a ridgy soil in Ephraim's territory.   -

In Eli's time the manifestation of the words of God was considered uncommon,hence precious. Eusebius pointed out the zone in his Onomasticon that Hiero - nymus translated later.   -

Probabily it was an ancient place full of market gar dens, particularly on the neighbouring slopes, as can - be deduced from a letter to Eustachium reminding - Shiloh under these gardens[36]. In the sanctuary ,which is the most important for Jews,take place nation's fun damental cerimonies and it's exercised a controll on - the cult. This was put into practice in a more spontan eous way elsewhere with different manners.   -

From the temple resound Hannah's words and her canticle[37] .The feasts become occasions to bring the - tradition to light;but she has a personal devotion ex - pressed in a meditation [38]. In her words,written with the style of psalms, there is joy coming from a salvation, exultation because she has had help by seeking comfort in God:"Strength is exalted in my God".   - Samuel will be granted custody to Eli, called governor by the Bible' s seventy interpreters. In Hannah' s - thanksgiving,the fortune is ascribed to Yahweh who - doesn't have equal if one looks at his sovereignty - and wonders[39] . It presents poetic features and topics by a style depending on the editorial phase. Elkanah, prophet's father,will go to the temple with humility - and deep reverence.His son was named Samuel be - cause,as the woman tells <for this child I prayed and the Lord has granted me the petition>.The boy grew up in the presence of the Lord, and won't make use of alcholic drinks .One can recognize the pattern that biblical authors also use for other exceptional events, for example Isaac's one. The principal elements speci fying the subject are the following [40]: the sacred,a - fear,a communication, prayer to obtain a sign, and fi- nally the accomplishment of what had been demand ed .In Shiloh the divine manifestation didn't happen owing to place's celebrity or because the temple was accomodating the ark of covenant;more than at holo causts and sacrifices Yahweh rejoices at the obedien-

ce and fidelity of Samuel. The ark was equipped with
four rings utilized to transport with poles. Famous are
the cherubim on it .As guardian geniuses for king's pa
laces they are also depicted on thresholds of mesopo
tamian buildings.They give solemnity to the ark,thro
ne of Yahweh. Cherubim appare during manifestati -
ons of God's glory together with hurricanes, earthqua
kes and fire. Shiloh's temple is only mentioned in the
Bible; pilgrims yearly went there for ritual feasts. The
archaeological evidences prove [41] that Shiloh was oc
cupied in the bronze age and flourished up to the i -
ron epoch . It loosed weight with the monarchy's be
ginning,but lived in splendour during Seleucids' time.
Hellenistic ceramics,country houses with bathroom, -
defensive wall,necropolis may be useful indications[42].
In the Chronicles' books it's asserted that Elkanah be
longs to a levite family;his stock coincides with that -
of Kohath,one of Levi's sons.   -
He was usually present at Shiloh owing to his family's
tradition. The levitic origin isn't explicit in the Samu -
el's books where his father is not linked to a priest's
context.It's worth to specify the Jewish expression for
a regular visit to Shiloh is <from time to time>.Psalms

will quote the temple as the place where God has his dwelling.Nevertheless,they were unfaithful in the sur roundings. God rejected Shiloh's tabernacle.Jeremiah will point out this event and punishment as a God's - judgment, because sons and dauthers were burned - out of Jerusalem's walls,in the high place of Tophet,in the valley of Hinnom's sons (Jer7).   -

The pagan practices: pouring out drink offerings and making cakes in honour of the queen of heaven.- The name high place,as a pagan area ,acquires a sym- bolic meaning with Jeremiah ,because Tophet is in a valley[43]. The abominations are denounced   mention- ing the innocent blood, foreigner gods,incense's offer ings to Baals, gods that Israelites have not known.   - Moreover it's stated one hasn't to oppress neither an alien,nor an orphan or the widow(Jer 7.6).   -
Eli's sons were wicked and died when the ark was - captured by Philistines. The sacred object won't re - turn to Shiloh; after a stop at Bet Shemesh it will re - main at Kiriath Jearim for years.Shiloh's feast, during which pilgrims climbed the sacred mount with flutes - and songs,was authorized owing to a holy and honor ed tradition.Notables were invited to banquets to eat sacrifices'meat[44] .   -

In Shiloh,the Ark is object of a sanctifying worship. As Lord's palace,the temple is his house existing before Solomon.We see Eli near to its door. Among the biblical traditions of Shiloh, one has the linen ephod for - priests, the vestment that Samuel weared.It tell us - that consacrated people belong to God from whom the blessings are originated; its decorations express - God's glory and the diadem a precious gift.   -

In a symbolic sense the ephod will represent saints' right works that God will appreciate with the prayers' perfume[45]. It's recorded a disease with its spreading among men and rodents owing to the action of an e - gyptian flea.The plague has mice as carriers[46].The description of the illness and its diffusion might have some importance for doctors.   -
Samuel had a house at Ramah,six miles north of Jerusalem . He went on a circuit, year by year,to places - such as Beth-El, Gilgal and Mizpah.   -
According to Hieronimus, Ramah's neibourhoods lay on Timnath's district; one is at west as respect to - Ephraim's soil[47].Mizpah will be a meeting's centre for Jews,very close to the Ramah's altar.A pagan cult flares up at Shechem,near the eastern entrance of the pass existing between the high peaks of Ebal and Gerizim. It was a passage to go westward through the - mountains of Efraim's region. Shiloh has a distance -

of 15 Km from Beth-El.; Dan will be chosen as capital
by Jeroboam for its position, important from the stra
tegic point of view. Excavations at Beersheba,where -
they used rudimentary arms made of bones and ivo–
ry,allowed to collect the stones of an ancient altar -
with horns.A snake,pagan engraving,as ornament on
it revealed the syncretism in the city[48].   -
   Shiloh is the promoting center of Yahweh's religion.
After the fall(Ps78.60) of the city,when the Philistines
prevailed over Israel,its inhabitans, pilgrims coming -
from Shechem and Samaria will go to the ruins of the
temple for peaceful offerings.   -
God had<delivered his power to captivity,his glory to
the hand of the foe(Ps78.61):in fact the ark is symbol
of his might and his splendor(Zondervan,p.880). One
already finds a secondary temple in Dan at the ark's
time.Jeroboam will build his altars at Beth-El.   -

Samuel prophet
With Samuel, one has a narrative concerning the ark.
It was obviously important, particularly in Jerusalem
to inform visitors[49]. Samuel called all Israel to Mizpah
for libations.They drew water and spread it out befo
re Yahweh , fastned on that day(1Sam7.6)and confes-
sed their sins.It's easy to think it was the tenth of the

7th month ,day of atonement(Lev16.29),unique occur-
rence of fasting. Hence,the prophet urged that they -
had to put away the hated worship of Baal. Such a na
me is referred to the canaanite god of the storm.   -

In local cults they worship another god adding an attibute to Baal .In Saul's prophetism we see an ecsta tic partecipation in a spiritual manifestation, pheno - menon appearing among Arabians too.The right trace is pointed out by the judge following the divine plan: < To obey is better than sacrifice, to heed than the fat of rams(1Sa15.22)>.   -

Another developed aspect is priesthood.Zadok is the priest of both David and Solomon.The zadokite des - cendants will have absolute authority after Josiah's re form. The coming of this king,foreshadowed since the time of Jeroboam's pagan altars, brings about idols' - destruction.Jerusalem[50] coordinated cult's ministers - for all the land.Priesthood's office follows vicissitudes depending on those of the ark.   -

Before Samuel story,patriarchs performed various functions pertaining to priests'control.   -
The priests have to get the word of Yahweh across to the people and to grant justice's order which origina - tes from the divine will.The theaching involves a set - of values and usages;and the priest inserts the sacred in the existence's reality. Samuel is put as communi -

ty's head .With his life,he marks transition's phase to prophets' period and represents their prototype.   -
Similar to Samuel story,under various aspects, is Sam son saga[51].   -

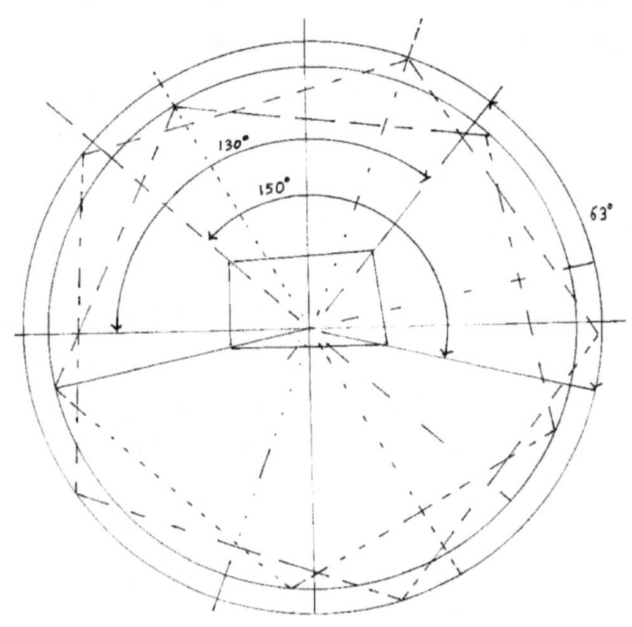

Fig.8- David's citadel(11th century B.C.).

Solomon's temple

The social organisation on the palestinian land is diffe
rentiated in 3000B.C. :if one works with metals in a -
district,we find ivory[52] elsewhere.To go to the capital
of Galilee it won't require many efforts:it is at the cen
ter of the occupied territory.   -
With Solomon begins the temple's construction and
it's allowed to be regularly present at sacrifices cele -
brating tradition's feasts. The patriarchs visited She -
chem,Hebron,Beersheba to see Yahweh.   -
About 1850B.C.,Abraham will buy the grave's area. At
the desert's limit Beersheba,where bedouins were living,offers a market to nomads(about 35 thousend at
present[53]).Salomon's temple was destroied as well as
the new rebuilt after the exile.On the large square of
the mount,sacred for Israelites,rises a mosque,adorn
ed with polychrome marbles[54] and Koran's inscripti -
ons.Before Israelites' time,Jebusites had been in Jeru
salem.At the end of the fourth millennium it towered
above the country's important roads because of its -
hills. Melchizedek is subjected to the Egyptian empi -
re. The known letters found at Tell-el-Amarna,200 mi
les from Cairo,contain information on this topic.   -
The inscriptions,with cuneiform letters ,report episodes involving Jews about 20[th]century B.C.- Joshua will

defeat Adonizedek. Yet,David is responsible for a grea
ter unity of the twelve tribes.   -

With care they build the temple, though an idola
trous cult remained for Tyre's influence. There will be
a shrine dedicated to Baal in Jerusalem.The city, with
thousands of nomads was secured on the hill[55] and -
fortified later.   -

Before the temple, whose bronzes are melted by
Tyre's workers and specialized artists,it's put a mag-
nificent bacin for ablutions. Beneath its base there -
are twelve bulls that will be removed in later epoch.
The architectural equilibrium obtained with hollows -
and overhangs requires an adequate ornamentation.-
As ark's building ,the temple is a place encouraging
the chosen people [56]and source of hope.In Qumran's
textes,the water's problem is fully discussed.One had
a wash at fountains put not less than 1375m.(= 2500
cubit)from the temple. They hadn't to be put at west
because of winds[57].At est the mounts stand out and
one hasn't to choose such a direction. At the other -
cardinal points they couldn't be seen at once. There
fore they are placed at N-W and at right of those lea-
ving the sanctuary(p.57).Temple's orientation is along
a line est-west,and its entrance is at est(p.52) .   -
The ark is covenant's sign;the sacred place proves -
the selection from Yahweh who will build a house for

David.Committed to Zadok with a splendid cerimony, it consisted of three main elements. After a vestibu le one entered the Holy Place(outer sanctuary20x40) and the Most Holy Place(Debir=inner sanctuary), cal- led Holy of Holies[20x20]because it housed the ark of covenant.This inner sanctuary was opened for the of- fice of the high priest once a year. It was completely kept in darkness (1Ki 8.12). Two cherubim, half the chamber high,were <guardians> of the Sinai tablets. They were made of wood covered with gold; had a certain resemblance to finds(of Syria and Phoenicia) with sphinx-like appearance.   -

For various peoples,the cherub is an angel, both counselor of god and defender for believers[58].   - Outside the vestibule,in addition to the bronze sea, o ne has an altar made of the same metal[59] with steps.- It's fixed the distinction between(power or) anointing of priests and that for kings. Sacrifices'minister is the priest, and one has to recall that even there be the   - ark, the place containing it isn't the sole for the divi - ne presence: Yahweh hears the prayer from his celes- tial dwelling. This concept precisely establishes what Israel had already understood during the clashes with Philistines:it rises the idea of spiritual cult.The temple itself is source of that grace,called cult's living water in subsequent epoch (Jn 7.38). For the building ,the -

stones were carried from Galilee to Mount Moriah, -
north-est of Jerusalem, close to the king's palace.   -
Adorned with refined style,using cedars, olive's doors
with streaks and numerous ornaments,it was conse -
crated by a feast with offering of 22 thousend oxen -
and 20 thousend rams [60]. It became place for discussi
ons, researches,education and important decisions.-
There was an inside bronze altar with four horns whe
re the rich offered animals,the poor cereals and do -
ves . Hiram had sent experts.Sidon's cedars and cedar
logs reached Israelites by rafts in exchange of oil.   -
During the reign of Solomon one had prosperity owing to king's wisdom. David had defeated antagonists
expanding the border of the tribes'alliance from out -
ermost north to Beesheba; it was a result stemming -
from Yahweh's benevolence for his people. Such an
attitude is expressed by psalms that, with poetic fea
tures,were used in royal palaces[61]and became a list -
of prayers after the exile.   -

Observations.
Introducing Palestine's temples ,it was also necessary to describe what one finds at Hazor,Gezer and Bi
blos.At Hazor and for Biblos(p.37) as well it's possible
to see stone's obstacles controlling sources of light -

during their motion(for example,sun,planets and so -
on). Actually, with a casted shadow on the ground at
Gezer one finds the latitude (of the city) needed for a
sundial.Such a device is also useful at Aijalon,with lit -
tle difference of time.   -
   Ark's temporary building(first temple)was located in
Shiloh,at 15 Km from Bethel.The boy Samuel born af
ter his mother's prayers was brought there for tem -
ple's service. As grown up he presides over penitenti-
al meetings in different territories. The captured ark ,
given back by Philistines, doesn't return to Shiloh.   -
Previously Jerusalem belonged to Gebusites. At the
time of David,in the city had been living thousends of
nomads. He expanded the border of the league's tribes from the extreme north to Beersheba.   -
Solomon will be the builder of the temple within Da -
vid's citadel (p.46). Famous are the cherubim guard -
ing the ark and involving a differentiation as respect
to analogous Mesopotamian figures.Source of grace,
which in subsequent epoches has been denominated
living water of the cult,is the temple.   -

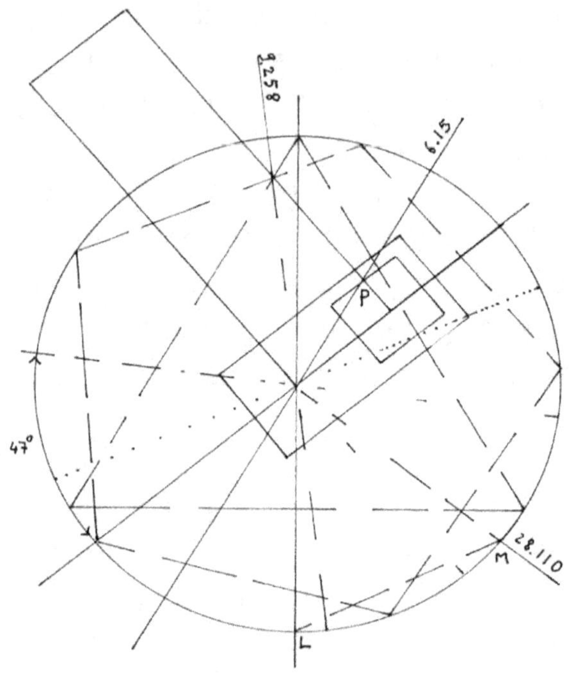

Scale   12° | 1.008
         13° | 1.092

Fig.9-P is the northern* entrance(Ez 8.16) of the tem-ple's inner court.   23°.5 =solstice(Romano,p.37);
47°=2(23°.5) .   As R. Bultmann[63] underlines,
< after the little girl was restored to life (by Jesus' miracle), the
report of the event spread throughout(Matt 9.26) >   -.
The spreading is handed on only from Matthew.   -
* The Zondervan NIV   Bible Commentary-Volume1: Old Testament - Zondervan Publishing House – 5300 Patterson SE –
Grand Rapid s- MI - 49530; 1954,p.1350 .   -

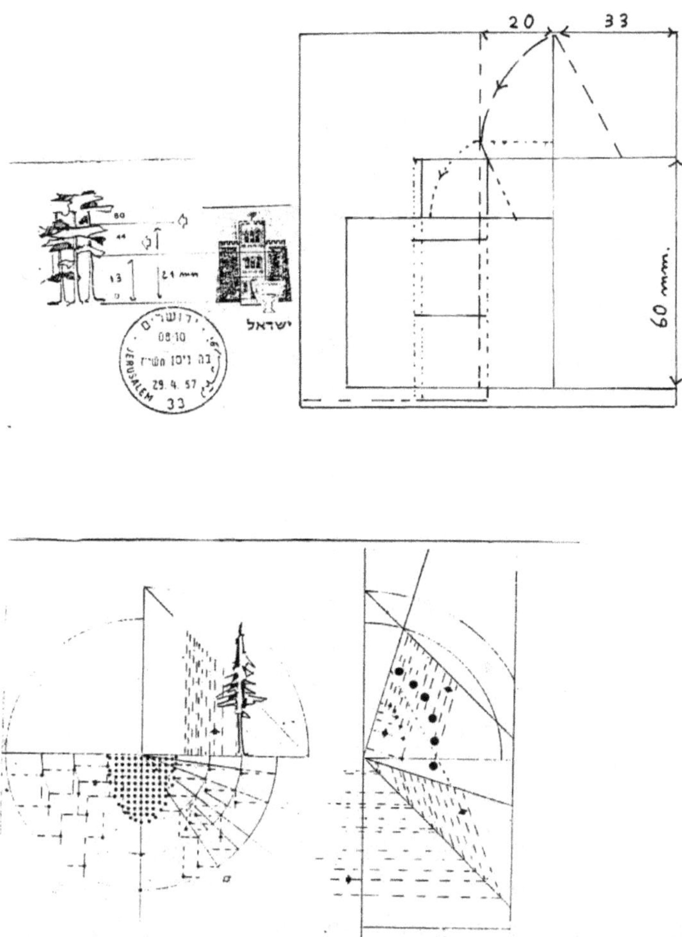

Fig.10-Above:stamp (1957) with temple's front. In the
lower part of the page: seven-branched golden ca<u>n</u>
dlestick.

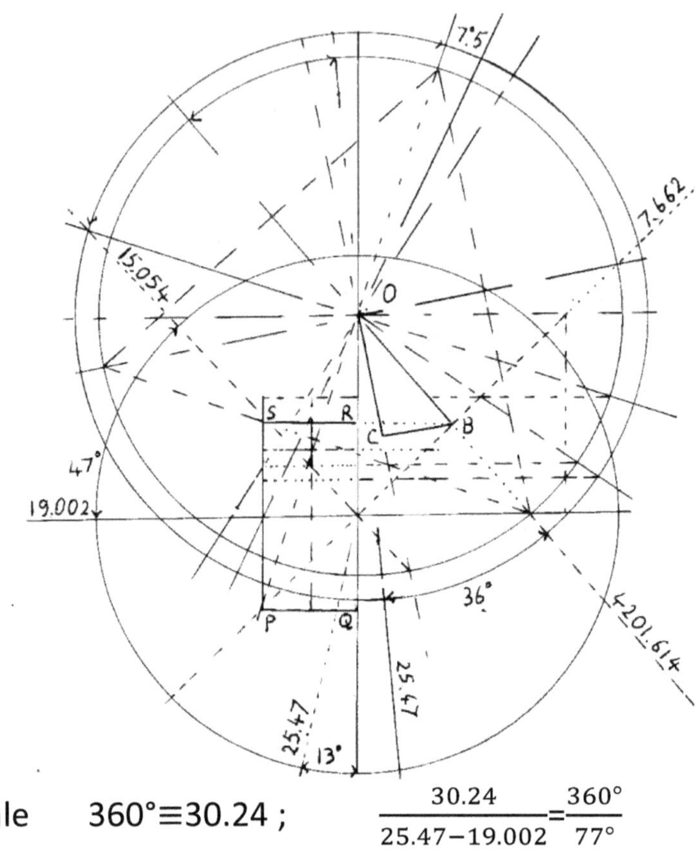

Scale   360°≡30.24 ; $\dfrac{30.24}{25.47-19.002}=\dfrac{360°}{77°}$

Fig.11- PQ=25mm. After the temple's destruction,the prestige passes to Jerusalem. B and C are two corners of David's citadel. OC ≡one of its walls. In prayer, three times a day David is mentioned as<messiah of **God's justice**>. PQRS≡ enlargement of Ezekiel's temple translated out of the citadel;Solomon's temple(p88) is in the upper half of this rect̲ angle.   **360(2547.9-89.4)=**

=29.53058025(29970.96544) (p.110;p.100) .   -

Ezekiel's temple[64]

Prophets' activity becomes important when the city's progress brings about an organized structure . - Then comes to sight a tradition,administration's pro - blems linked to the growth of the urban agglomerations rise and various social ranks must be supported owing to a blind individualism .   -

A discipline for priests arises.Zadok's successors attend to the altar and must wear flax cassoks[65].   - In previous times, temple and palace were contempo rarily situated in the same complex as happened for o ther peoples.In Egypt there is confusion: the tombs of pharahos(representative of the creator-God) lay near the temple(Bernhard W.Anderson,p.43).
Ezekiel's measures are well defined to reorganize the cult. With the dispersion, the State wasn't successful in carrying out the religious mission difficult in accom plishment, and Jeremiah predicts a new religion with heart's transformation. Ezekiel will be more the the ologian.During Solomon's reign,the ark is sacred; but holiness'prestige passes to the temple.
With no comprehensive descriptions,it's was absolutely needed a reorganisation with more details;the re former's one which appeares to be adequate.   -

The levites <had charge of the work in the temple (1Ch 23.4)>,intervening when burnt offerings were

presented to the Lord (1 Ch 23.31).The priest was at
nation's center.One had to eradicate violence ,robbe
ry; to check balances and weights. It disappeared the
figure like Melchizedek who was king and priest at -
the same time. A sovereign is a layman joining in ceri
monies and contributing toward cult promotion[66].   -
People's life receives stimulus with meetings and because the temple is stable structure.   -
Jerusalem's destruction had brought forth the sinago
gue's reality;therefore both words' cult which prophe
ts carried out and sacrifices offered by priests.The hu
man values are saved and handed on.   -

Torah's meditation yields a progress safeguarding
the people from mistakes. Through psalms one will -
praise both wisdom and brotherhood;examples invite
to emulation.In spite of a messianic vision,the Jewish
system is for an ethnic group without purposes of uni
versal expansion. Through the activities and instituti -
ons originating from the religious mentality it will be
possible to understand the change produced by Eze -
kiel . Therefore one can't neglect anything regarding
the religion.Thus one will pay attention to the liturgy
and every religious phenomenology included in pilgri
mages, sacrifices,gatherings and prayers. The eschato
logic aspect is another element referring to the hope.
As reference point of the religious ideal remains the

sanctuary. Ezekiel has a project extending over subse
quent generations:his purpose is effective.After the e
xile one notes his influence on the prophets and the -
refore the richness of his message.It's amplified what
Jeremiah preconized. Being different the contexts in
which these prophets lived,their problems change. -
Ezekiel plays on religious feelings and trust in God ba
sed on direct dialogue; Jeremiah wants the hearts be
purified,and the temple will express his program.   -
When Solomon was alive the city was in the south of
the temple and there weren't needed three entran -
ces. After the 8ᵗʰcentury,a city's enlargement will be -
at west on the hill and in the valley up to closeness to
Mount Ofel. With the **western** door the inhabitans -
will be allowed to enter the temple without turning
round the walls. It is a royal sanctuary unlike others -
such as that of Bethel.Vainly Jeroboam with northern
temples will try to modify that religious feeling inspi -
red from Jerusalem. In his restoration Ezekiel doesn't
leave anything to chance. In the biblical description -
the surfaces,by him well defined,catch the eye.   -

    This has a goal,particular and innovative as res-
pect to the narrations concerning the temple of the -
Kings' books[67].In the Holy Place(20 x40) there was the
wooden altar of incense. The Most Holy Place(= Inner
Sanctuary) illuminated once a year recalls the glory -

and mystery[68] of the Lord .The cherubim introduced -
the faithful to the sacred,presenting worshippers to -
God. However the Jewish concept is richer than that -
existing elsewhere.Faith's monotheism confirms it. -
Moreover they are alive. This is a revelation of the -
chosen people. Similar beings are also found in Babylonian pagan figurations.

Hence,Ezekiel has to dwell upon a precise character
ization involving elements relevant to angels,God and
man. The celestial spirits, in psalms,are compared to -
human beings; so we can guess the specifications a -
bout cherubim implicate a representation of the su -
pernatural,therefore a theology.

Authentic religiousness had been prepared through
Solomon's wise attitude.

Now, considerations are more articulate and the holi
ness' attribute is underlined. In Ezekiel the feelings re
flect a tormented period and foreshadow all the vi -
sions apocalyptic in character.One wants to estabilish
the independence between profane things and the sa
cred.For this purpose use is made of different depart
ments conceived as prolongations of Solomon's struc
tures. Temple, in its etymology,means house of the -
sanctuary. Therefore it emphasizes holiness and pre -
sence of the sacred[69]. In its stable building,and earlier
in the desert,one went before the ark to be at the pre

sence of Yahweh.With psalm 47 one recalls the appel
lation Elyon (=the most High) as Yahweh's attribute. -
El-Elyon is also invoked by preexistent peoples,in the
occupied land. The monarchy will assure nation's uni
ty.Allowing and promoting the cult will be the guiding
principles of it. David has the spirit of Yahweh,whose
protection is verified during the dramatic moments -
of battles.After Philistines came Assyrians.At the time
of Isaiah,hence with Hezekiah and Manasseh, the Assyrian domination will carry its cults among Jews. Ni -
neveh's destruction indicated the decline.
Afterwards,one has the struggle between Egyptians -
and Babylonians for the supremacy. A revolt against -
the last people will cause temple's destruction. In the
Leviticus,the ark with its cover is an instrument opera
ting by virtue of the blood shed on it for the forgive -
ness of sins. From the juridical regulations of such a -
book, in Ezekiel's reform one goes on with a widening
of the spiritual dimension. The blood consacrates the
altar.A priest will offer the holocausts and pacific sa -
crifices on it. In other words, people's sacrifices are -
given by oblation and holocaust.   -

The ideas about the sacred.

Cult's main acts are the prayer,offering and sacrifi -
ces.While Israel's God is transcendent,the Babylonian
gods[70] are sublimated transposition of the human es -
sence. They have more power and science as respect
to men.Hence,different is the attitude towards the di
vinity. At Babylon one addresses gods to avoid wrath
and obtain favors. For Jews it doesn't suffice an out -
ward manifestation like to make offerings.There must
be dispositions of the soul respecting spiritual values
such as justice and virtues.   -

      From Proverbs,ascribed to Solomon,we deduce
that, for Israelites, king's greatness is to wisely percei
ve good causes.He is an enlightened ruler stating the
law.Hammurabi's code is the composition containing
customs,drastic innovations and amendments[71], pre -
scribes the strong mustn't oppress the weak.   -
It actually suggests a rules'collection that will be modified in subsequent epochs. After a damage,it was -
allowed to retaliate on individual human being.   -
Privileges were bestowed too. Deat penalty could be
inflicted and this will require further alterations to a-
void it.In Jerusalem one doesn't ascribe a celestial ori
gin to the temple; it's lacking for the temple a myth -
echoing a cosmogonic event ,as happens in Sumerian
and Babylonian legends.David conceives the temple's

idea and decides where it must be built; Solomon car
ried out the plan, but the gift of the divine presence
comes from heaven.Neither David nor Solomon or rite,or prophet can impose an automatic santification.
God is sovereign and grants grace[72].

Summarizing, two are the dominant convinctions,
that of transcendence and the other of history. They
are particularly implicit in the considerations about -
the temple. As for the creation, Eusebius asserts that
with Jews one gets a rational judgment and pious
application to study the universe's nature:
<..And so...they saw that
the bodies' basic elements,earth,water,
air,fire,which,according to what they had realized
 compound the universe,weren't
 more than sun,moon and celestial bodies,gods;
 but God's works>[73].
In Ezekiel's reform,temple's measures will furnish ra -
tios similar to those found in **Egypt** (p.118) and for -
Greeks.At west,there is the Most Holy Place (=Debir).
 This doesn't happen for buildings dedicated to Mar -
duk and generally for analogous edifices.In Israel Mel
chizedek is both king and priest. One has then distance from Babylonian and Egyptian customs. Israel's a-
nointed is endowed with Lord's spirit,but the person
can't be deified; his task is to create conditions in fa-

vor of the cult.Moreover he must look after the sanc
tuary with practical help[74]. The priest is called to holi
ness,executes blood's sprinklings,offers incense, obla
tions,sacrifices.   -

Second temple.

 Cyrus, king of Persia, after his victory over Babylon -
ians wrote his famous edict;then in 537 B.C.; Zerubba
bel and Joshua will guide the return to Jerusalem.   -
With skilful political action,the vessels that Nebucha -
dnezzar had carried away from Lord's house were gi
ven back to the Israelites.Maybe the Zoroastrian religion[75] had some influence on this decision.
There will be need of some years before the sanctuary be rebuilt. Faithfuls' care prepares the restoration
of the temple's functions making use[76]of the golden
altar for incense,of the table for loaves' offerings and
of the candlestick. An explanation of the theological
reasons involved in these furnishings is given by Philo
of Alexandria:they want to be a symbol to sing Lord's
praises for what we find in the universe and for hu -
man works[77].No component of the world can be accu
sed of ingratitude.The candlestick implies the thanksgiving for all what we observe in the sky. It refers to
the supernatural. The table means praise given by -

workers;the remaining element shows gratitude for -
what is free gift. They will be arranged as follows: in
the Holy Place there is the altar,on its right-hand side
the candlestick, on the left-hand side the offerings. -
When Pompeus came in Jerusalem(63A.D.) the front
of the temple was decorated with golden crowns.The
eagle near them was removed by a Jew. The vestibu
bule, also called bit hilani, is typical construction with
columns. This same expression will designate a whole
building with arcade. A first example of it is found -
at Alalakh[78].As accessory there is often Olives' Mount
for didactic purpose: according to the tradition, the -
priest looked at temple's entrance from an eastern -
superelevate zone laying across from the Kidron,whi-
le animals were immolated. The building near the al
tar of stone as high as that of Solomon was surround
ed by thick walls.The coreography,hieratic looks, inside dynamism,and strictness characterize the whole. -
Psalms give information on liturgy too. In Jerusalem -
the works supervised by Zerubbabel were enthusias
tically welcomed and there was the dedication when
Darius reigned. Disastrous subsequent times were at
the height with a profanation of Antiochus ,surnamed
Epiphane[79].Long after temple's rebuilding,a Jew des -
cribes the effect of the sacred as follows:

<Blessed who saw the beauty's splendour of its great
ness and all the acts of its power and force.
Lucky who hopes and tries to see it . Could you wish
the temple be soon rebuilt today,in order that our -
heart[80] be rejoicing and our eyes admiring>.   -
Haggai's words exorting Zerubbabel and faith's char -
acter among the Jews [81]are thought back.   -
The leader will be the olive tree depicted with Joshua
near the candlestick whose lamps represent the eyes
of God scanning all the Earth.

   The second temple reached higher height than the
former by enterprice of Herod who wanted peace in
the territories governed by him.   -
Such a temple[always considered the 2nd temple with
embellishment(F.Spadafora[135] -p.587)] is the most hu
ge and splendid among his several works[82]. The Gos -
pel quotes the impression excited at the sight of its -
blocks of stone. A governor doesn't ignore they ga -
ther there for sacrifices. Well defined dimensions has
the square on which the temple stands.Out of the zo
ne set aside for priests' office one finds an external
part reserved for men.Women have to take part in -
the execution of rites from away .   -
Pagans frequented the remainder of the plaza .Very -
beautiful are the colonnades leaning on the four peri
metric walls ;two of them are real arcades with co -

lumns 12.5 metres high. At south one doesn't have -
two rows of columns but four. The estern of them is -
called after Solomon;from its side one could make for
Bethany passing through the relative door. The begin
ning of the Christian era will always consider the tem
ple as important point of reference.In Eusebius we ha
ve metaphoras concerning the bishop Paulinus of Ty-
re(4th century)known for constructions and restorati
ons[83]. Rethoric's laws permeate through the words, -
but leave behind temple's extraordinary magnificen-
ce[84].It's already time in which the Church is indicated
as Solomon's new temple.

Of Herod's temple,destroyed (70 A.D.) by Titus it
doesn't remain more than a wall. In 1948,when the Is
rael's State had been constituted,one couldn't even -
be near this wall,because it belonged to Jordanians. -
The impediment was removed after the six days' war
(1967). Since 1897 the Zionist movement grouped -
Jews in order to found the State.In subsequent peri-
ods the problem of scattered Jews in world's regions
will come to the forefront. In particular one has to recall a Paris' attempt(1980)which shaked the public o -
pinion. Today,Jerusalem's new buildings distinguish -
themselves by delicate colours. The recalls of cult's -
ministers,those of Jews,Christians and Muslims echo
in the city,sacred for three peoples.From Mount Ofel,

at 150 meters above the Kidron, dawn's appearance was announced by sentry. Death penalty prevented pagans from crossing the temple's threshold.   -

Considerations on the first part
The phases of Ezekiel's restoration,for a reconstituti-on of the whole sacred in places where it was more difficult to win idolatry,have been inspected in chro-nologic order.It was therefore essential to look also at previous epoches with paragraphes clarifying pro-blems .Colunmns put as embellishment,furnishing -such as the candlestick and the other objects used du-ring the rites enrich the precious heritage of the cho-sen people from both the material and symbolic po -int of view. All the drawings have been made on pur -pose to catch essential features of the data. Each of -them represents contents derived from the sources interpreted in objective critical way. They collectively prelude further researches both on Israel's wisdom – (as it's wished through Von Rad's guiding lines) and -the quantification's processes then known[85] .   -
In the appendix we show elements of a table,taken from pharahos' arithmetic which actually proves to be a pattern concerning the moon[86]. The theoretical -

plan of Ezekiel's temple represents a structural evolu
tion of Solomon's one,including it.Such a complex has
been exactly superimposed on stars' connections in -
fig.15,obtained by the use of a NASA's photograph[87]
in a spatial domain comprising eight galaxies,among
which Andromeda,and globular clusters.If one recalls
temple's history one finds that it derives from contri-
butions of skilled men.We aren't surprised if the human ability adoptes a type reflecting a criterion such
as functionality,reducing its necessary dynamics to -
what is fundamental,respecting what a sacred place
requires for celebrations and architectural procedu -
res. All what concernes the temple has been inserted
in a christian way within a both theologic and teleo -
logic frame. This sanctuary,as dynamically structured
unity,implies a future perspective.   -
To examin it with geometry isn't important. Actually,
Yahweh's dwelling respects those conditions forming
the enchantment of the environment. It suggests the
idea of covenant ,and it's truly a sign of Creator's prai
se.   -

David and Abraham's God

After Israelites' gathering ,the ark is transferred from Kiriath-Jearim(1Ch 13.5) to the house of Obed-edom and remains there some months.Uzzah who guided - the wagon upon which was the Ark died before reach ing such a place.The area where the temple will be - built was decided when David saw an angel by the - threshing floor of Araunah(2 Sa 24.17).Hence a king - will be able to offer appreciated holocausts.   - Pestilence's danger is avoided(1Sa 5).In Ezekiel,a holy place is called miqdas[88],but it isn't identical to a buil - ding. David prepares different furnishings in order to adorn the housing of Yahweh. In a prayer of the dedi cation we grasp characteristic elements of the Jewish religion. God is holy;one addresses him because invo cation,plea and thanksgiving are listened. The Lord - isn't depicted with images;it's forbidden from Moses' law.It won't be possible to have anthropomorphic re presentations imitating other cults.

Famous is the Asherah, symbol of a female deity who se wooden exemplars are found in resorts populated with Baals. For Israelites is Yahweh who, according to his inscrutable decision,takes the initiative to save,in a historic dynamism. As it will be clear later,the dialo gue between God and men starts with the covenant,- regularly renewed through a faith's confession.The idolatry extirpation is imposed owing to God's holi - ness. We can think of the temple comparing it to a - house; this is an idea common enough among peo - ples[89]in that epoch. Hence, if Israelites mention an E - kal (as inner part of it),such a term exists in akkadian language too. Atonement's sacrifices and propitiati - on's rites were carried out at the temple.

Pagan cult is noticed in Palestine; one finds it at He - bron and Dan.As for the temple structure,it is various enough. Solomon' sanctuary isn't like a ziggurat,of - course.Generally,in different resorts and also for Me- sopotamian complexes one finds an enclosure,whose inner space is sacred, comprising a building where a - god resides or goes.In Uruk we have the most ancient testimonies[90](p.113).   -

In comparison between Egypt and Arabia one notices similarities; they refer only to 3000B.C. (as De Vaux - mantains it) and therefore one has to study towers - with various floors afterwards.

The sacred places

In the Pentateuch and the subsequent biblical tradition there is the continous memory of the lands visited
by patriarchs.In every document one must see if the -
hermeneutic sign constituting the salvation plan is -
continous. After a first mention of Noah, the just man
who doesn' fall in to God's wrath before flood, and -
who sees a covenant   with the almighty God,we find
Abraham's faith. Trustful in the Most High,he leaves -
Ur to come in the promised land. Shechem,Hebron,-
Mamre,places that he visited, have been areas of different divine manifestations openly marked with the
altars put from time to time in order to invoke Lord's
name.   -

Altars and Shechem's sanctuary

Patriarchs had worshipped God who revealed him -
self to them before they came in Shiloh. With Abra -
ham's vicissitudes one will meet two altars put by -
him in Shechem and later at a resort laying between-
Bethel and Ai(Ge 13.14). The Canaanite area, ruled by
pagans,was wide in relation to its population's densi -
ty (Ge 34.21).There were actually religious places in -
Shechem(Ge12.6).Before buying the Shechem's fields
where he put an altar to El(=Israel's God),Jacob had -

build a house at Succot(Ge 33.17). This happened af -
ter his reconciliation with Esau. Bethel's foundation -
had already been carried out.   -

   The necessity to specify the indispensable instructi
ons to approach the altar rises[91] with Aaron. In the -
Deuteronomy(with Joshua)one can deduce there had
been altar's use on the Ebal(Dt 27)(where one read -
the blessings).
The place was not far from Shechem which is situa -
ted between this mount and the Gerizim. Moreover
there is instruction about what must be read during -
the feasts. So it's assured that a rite be repeated.   -
Abraham had found a sacred place(=maqom) gifted -
with oracle.   -

Bethel

In Jacob's time pagan deities are still in Shechem. The
patriarch goes to a place with more altitude(Ge 35.1)
and underneath the oak are buried all the alien gods
and earrings(Ge35). At Bethel(Luz) it's put a Jewish al
tar and such a place is then called El-Bethel (Ge 35.8).
Jacob moves away from this camp going near Bethle-
hem(=Ephrath)). At Mamre (Kiriath - Arba) his father -
Isaac is waiting for him in a zone where Abraham had

been already entitled to the pastures and settle with an altar(Ge13.18). Machpelah,occupied by Hittites, is in the east of Mamre.Also famous is Beersheba's dis - trict.Here,after an agreement with Philistines' leaders Isaac drills a well.Today, owing to this pact, the city is named after the well (Shibah).   -

Bamoth and cult

Abraham goes away from Ur of the Chaldees, but the Babylonian region is linked to the temple's painful vi - cissitudes till its destruction. Therefore it will be use - ful to specify different Mesopotamian characteristics, regarding cult especially.In Babylon one believes in se veral deities.The religious conceptions make use of i - mages as signs of the divine presence. Moreover one notes a cosmic background. In a sacred enclosure it's put a complex comprising the temple.It's made use of a structure of pyramidal shape and of its alterations. - The ziggurats are still somewhat mysterious.They re - veal the concept of a god going towards men in tem ples close to the celestial residence. Uruk's conside ration will furnish one of the key motives of the am - ple development of the sacred Mesopotamian sanc - tuaries. One must still discover what one wants to re- present or reveal also by the engravings and the figu - rative arts. In the geometric design an idea can be syn thesized in the lines.In Uruk, for example,use is made of an angle through 13°(for astronomycal reasons). - The most interesting idea in the religious context is that of the god who helps.It's something shared with pious Jews and it will appear during the apocalyptic period more explcitly.   -

In short,in the cosmological data,in the attributes of the respective supreme beings[92], elaboration of all - what was referred to seasons and celestial bodies , a bout 4000 years B.C.it was outlined belief in a god tur ning his action to men.   -

Idolatrous cult

Bamah is a term[93] which in various contexts means - ridge. Actually the bamoth are largely spread and dis tributed also in level planes.In them one finds Ashe - roth and steles too. The most important of these high places in Palestine is Gibeon. The foundations as holy places, where the name of the Lord is invoked, neces- sarily occur in resorts enclosed in areas where pagan peoples are installed. The institution of a ritual is im plied in Jacob's words: he speaks about the tenth - part.Hence it comes into sight a zones' grouping whe- re the syncretism is one of the outstanding features. - It's present at Mamre .Here there was a market and - the way of mention in the Bible shows purpose of the

redaction,according to De Vaux, to give scanty importance to a center not controlled from the institutional
point of view.During the history we find a semantic e-
volution for bamah: it will acquire a negative colour. -

Observations

We followed the foundation of the Israelite first sanc-
tuaries. Owing to the other communities living in that
time,it's attached importance to consider Babylonian
monuments and particularly Uruk's ziggurat,because
it has a simple shape as respect to similar towers of -
subsequent epoches. While one finds the concept of
the god addressing men, and this is already clear with
Babylon temples,it's at the same time specified a pos
sible cosmic function(p.113)of these complexes.   -
The list of the sacred places displaies historical memo
ries:one steadily keeps an eye on cult's expansion in -
alien land.Yahweh's covenant and the altar sign inclu
de a revelation's idea;besides they concern temple's
theology.The commandments excluding any anthro-
pomorphism give a shake to numerous myths,and in
troduce to what the new covenant suggests.   -

Temple's theology

We want to see the meaning implied in temple's vicissitudes. At first sight it could be surprising a devotion persisting although the reigns were split after So
lomon and even with the destruction of the temple.-
1)Concept of covenant

In people's history,we see appeal to the transcen
dent interlacing with human episodes continously. To
Yahweh's cult,present with tent in the desert and la -
ter in the monarchic temple,there are associated con
crete elements enriching the redeeming history: the
gift of Yahweh's spirit is that of human freedom. An
undertaking' failure is often seen as punishment for
transgression of the divine instructions.This is parti -
cularly verified in temple's history. However the salva
tion is a good that can be invoked in prayer(1Ki 8).   -
There is gradually widening of the presence concept,-
because it takes possession of that spirituality recog-
nizing the transcendence.In the temple one celebra -
tes God and his name. The Earth takes in Yahweh's -
throne,but the whole Jerusalem can't contain the uni
que God of Heaven.The finitude[94],opposite of what is
transcendent opens the way for the speech of both
eschatologic fullness and God's fidelity.The < berith>
is a pact with juridical obbligations[95] assured from -
that reality which is the holiness of Yahweh. Cove -

nant is a certainty of grace and peace sealed with -
blood.Cause of unity and sign of the covenant is the -
temple. If the prophets critisized the warped vision -
that people had about the sacred, it's a symptom of
the necessity to face up to every cult's act with sound
realism. God shows himself with benevolence,but -
conceals a totality which remains obscure for men. -
The reason of the darkness in the Most Holy Place is -
based on this idea. On the other hand, God's holiness
calls for caution to properly go near the Ark.   -

          After specific sacrifices the priest will be able to
come in the Debir, with incensory and incense[96] ,in
order to beseech the Lord.   -

2)Chosen people

With David's coming ,Jerusalem is most of all celebra-
ted because it's privileged place for the Ark .   -
This is recognized as sign of the grace granted for the
people.The importance of the temple is more under -
stood after the exile,in fact one proceeds to built it
again. With Hezekiah a protection's sense grows be -
cause work of Yahweh is this king's salvation.   -

          With different temple's destructions this God's -
house is thought as a place where his manifestations
really happen. Israelites' choice begins with Lord's covenant. If we tell people,we denote folks convoked -
not owing to a philosophy or ideology of a common -

civilization. The grace reveals itself through the covenant and choice. It 's found implicitily expressed in -
Old Testament's terms [97] such as esed[covenant faith
fulness, mercy, benevolence (Ps33.22) ,steadfast lo -
ve], hen [supplication (Ps130)], emet [faithfulness -
(Ps85)],rahamin [compassion which is dominant incli
nation to forgiveness (Ps103),based on the fact that a
man was created with spiritual image].   -

The word covenant implies that people's concept
has been widened: it suggests a transfiguring commu
nion whose visible sign is priests' reign. The covenant
is the starting, but all is preparation,tension between
what exists now and the future. Covenant(=foedus) is
actually a merciful pact[98], renewed with sprinkling of
blood on the ark,first of all at est, according to the Fa-
thers.So they reminded the justice's Sun is coming. -
Yahweh loves his people that should be like a bride -
shining for good deeds(Rev 19).   -

David wears the ephod,sign of wealth stemming
from the divine choice; however it mustn't degenera
te in external sign which drag somebody to idolatry.
With priesthood, the choice reachs an apex perceived
through the revelations of both the letter to Hebrews
and the Apocalypse. One has a summary of temple's
theology in them. Choice doesn't involve only a Yah -
weh's protection for a people that has to defend it -

self from nations' traps. In fact there is a hope both
in eschatologic perspectives and in the possibility to
benefit from the goods of the earthly city.   -
Presence and faithfulness of Yahweh imply that the
salvation has begun,though not realized in a full way.
This concept is retaken during the apostolic time and
from Church's Fathers. It will be clear that the cove -
nant has to follow the spirit's principle when it will be
spoken of Israel's remnant,and about implicit judge -
ment implied in Yahweh's appeals[99].One must exclu -
de either every allusion to nationalism like that existing in David's time or a covenant relative to an earthly reign. God of covenant is the God of Abraham, patriarch living in the spirit of that pact stipulated with
Noah and also inscribed in nature.In liturgic expressi -
ons we note what is characteristic of the Jewish faith:
in sacrifices it doesn't matter the material aspect that
could be simple formalism. One has to note superfine
flour was offered by Abraham,while different is Lot's
idea .Hence there is a path in the variedness of the -
spiritual gifts .A sacrifice has to be measured on the -
grounds of the inner dimension[100].   -
   The Fathers have also interpreted the pacific offerings such as incense ,onix ,storax and galbanum in a
mystic meaning. The vestments that the main priest -
weared become expression for them of the various -

virtues a soul should have in order to hold a dialogue with the supreme God[101]. Here we find a revelation's concept according to which a mediator is needed:so - it's open the way for the christological view of Spirit's expectation. The announced salvation isn't exclusive gift.The covenant allots Israel a role among nations - that can't be mixed up with what the Jewish messia - nism calls for. Besides,also with Tobit is revealed that plan of salvation inviting all peoples to admire the - splendour of Yahweh's glory .   -

The salvation's plan is one of the hermeneutic ca tegories able to guide a just interpretation of the Jew ish history,and coupled with a right anthropology, per mits to avoid the negative aspects of the apocalyptic currents . Isralel's choice wants to remind men that - they have a dignity springing from their partecipation in God's holiness[102]. Hence every theory relative to a destiny conditioning history is avoided. God's praise - doesn't imply an attempt to restrict the consciences, but the purpose to introduce that salvation's myste- ry whose result is to put men in their more authen - tic dimension. If the covenant,from Zion's theocratic vision, perfects itself in the proclamation of God's so- vereignty, it's by no means estabilished a contrast be

tween men's freedom and a fictious being. Moreover ethics won't only be a reasonable convention. It's -
promise to obtain a free indescribable gift which is -
true presence of a men's (onthological) strengthening owing to spiritual progress .   -

Temple

The description of the sanctuary ,of the accessories -
and David's accurate preparations are found with va -
rious changes in the books of Samuel,of the Kings and of Chronicles. The temple is important for the people; the temple's institution for which a king supplies material and human elements.However it's through pro
phets' work, who took care of the divine world and -
practical observance of the celestial statutes,that the covenant with Yahweh could direct the moral behavi
or of the people and nation.Also the Mishna speaks -
about the second temple modified by Herod: there -
were nine years and a half of main work.   -
Starting in the 4th year of Solomon's reign,seven years were needed to finish the first temple.For it there is a structural system to guarantee stability: one has com
bined use of wood and stones. With the exception of

the front,on all sides it has a triple layer,that is, that -
part provided with trapdoors and winding staircases
almost certainly[104].The cherubim near the ark reflect
a symbology which speaks of the virtues and operati -
ons of these beings. In them it's deposited divine wis
dom and,as they belong to a hierarchy, give example
of continous adhesion to God's similarity carried out
through his imitation[105].

Solomon isn't directly involved in the cult after the
temple dedication. One mustn't attribute a political -
meaning to the cult's foundation in a peculiar place:
the religion in Israel emploies rules valid both for the
king and people. In the concepts of ark and cloud, by
which Lord's glory has filled the sacred place,is implicit that of dwelling ,dense of meaning for Jews.   -
As the priestly tradition observes,one can't limit one
self to a representation containing a simple tent or -
material temple . When it was constructed,a cloud -
covered it and was a guide in the displacements[106] -
(Nm 9). The temple is a cult's place owing to the pre
sence of the ark whose covering has an important -
function.Origene refers of the blood's sprinkling on
it on the day of atonement;a rite interpreted as fore

shadowing of the redemption granted by Christ, the priest able to intercede for men in the celestial Jerusalem.

As we shall see,these ideas constitute the apex of temple's theology,and as it's well known, are taken in to account both in the Apocalypse and letter to the Hebrews. The cherubim mark Yahweh's throne, but ark and its covering don't furnish a repository able to limit his omnipotence. Prophets'words, psalms' con tent show more explcitily these themes;and it was ne cessary in a vision of the sacred having its foundation in revelation's God.The cherubim are beyond the veil and have the benefit of the higest inner illumination.The matter of which they are made wants to suggest the preciousness of their closeness to God,for they are replete with a wisdom that they have to transmit. In this virtue they are generous[107]. Thus one wants to give prominence to a gift, the perfection of a spiritual image bestowed by the Creator.In the cherub one can't find any darkening. At the presence of the angels it's inspired the concept of God's imitation con sisting in a continous adherence to God's similarity impressed in them.Cherubim belong to one of the le

vels of the angelic hierarchy; therefore they also rein
force a concept of order understood as closeness to
the Creator. Angels' representation implies underli -
ning that their knowledge is more resemblant to that
of the supreme Wisdom.   -
Both at the tent and ark,visitors have to mind the gra
ce .Other figures could confuse the ideas in an envir -
onment refusing anthroporphism.

Temple before the exile

Ahead of its destruction which happened in 587 B.C. -
caused by Nebuchadnezzar,the temple undergoes al -
ternate vicissitudes depending on leaders' religious
ness. At reign's division we see explicitily how differ -
ent are the attitudes of the kings in the sacred field. -
In the span of 400 years,beginning with Solomon,one
has that wise guides take turn with irreligious.   -
The prophecy of Ahijah the Shilonite,directed to Jero
boam(1Ki 12.15) comes true. The kings designated as
pious in the Chronicles are few:Asa,Hezekiah,Josiah -
distinguish themselves by their reformations. The -
last of these kings tires himself personally demolish-
ing sacred poles erected in sun's honour (2Ch 34.3)

and making dust of carved and cast images .Actually
he didn't restrict himself to a temple's restoration ,
but controlled the most hidden places of Manasseh,
Ephraim, Simeon, reaching Naphtali's boundaries .In
the Chronicles one hints at the constructional thecni-
que for which use is made of wooden planks to give
more elasticity and safety .In fact this prodedure was
utilized both for connections and girders of temple's
buildings (2Ch 34.11). Previously Hezekiah had been
compelled to take precious objects from the temple -
in order to pay a tribute. The Bible speaks about all -
the works necessary on the mount for temple's pla
cing. Several details were brought to light during re -
cent excavations;thus it could be widened our reflec -
tion about the temple. It stood as wonderful dwelling
and was put at higher level as respect to the external
area having the colonnade on the boundary.   -

   Ezekiel doubles ekal's measure ,which was 20 for
Solomon'temple(Ez 41.2).There had already been the
irreligiousness of Ahaz .He made molten idols,passed
his sons through fire (2Ch28.2) , although heir of Jo -
tham who had been attentive to the temple.
Ahaz even plundered the temple to give tribute to As

syrians,and in his distress sacrificed to alien gods, to -
get help from them.A pagan altar replaced that used
for sacrifices and was changed the known configura-
tion chosen originally for the bronze sea.   -

  With Jehosaphat,king of Judah,sacrifices and incen
se's offerings were still made on the high places al
though his father Asa had tried to do away with -
them in a reformation carried out between 910 -870
B.C (2Ch 14.2). An action of Asa is very meaningful :
he removed his mother from office of queen because
she had made an image for Asherah (2Ch15).   -
After him Jehosaphat will be famous for his religiousness (2Ch17.3). He took an interest in extention work
for temple's structures. Thus,after Solomon,a note -
worthy toil is caused from the sanctuary's presence.
Foreigner rulers change. With Jeroboam one has an
Egyptians' first invasion.Suffering from hostilities Asa
was obliged to call for help from Assyrians .There will
be an attempt to usurp the throne,with Athaliah ca -
se. Once more,in this episode one catches a   glimpse
of temple's effectiveness:during the life of the priest
Jehoiada,his wife stole Joash away from among those
children that were about to be killed. The coming of -

Neco from Egypt to fight Jews at Carchemish on the -
Euphrates (2Ch35.20) proves situation's instability.   -
Josiah didn't listen Neco's proposals; then,in a battle ,
Egyptian archers shot him, bringing about his death la
ter. With one of his two sons who reign afterward,on
the Jewish land will intervene Nebuchadnezzar. Jeho-
iakim,made king by Neco(2Ch36),rebelled in vain.   -
Jehoiakin,his son,succeded him but was deported to
Babylon in a captivity for 37 years (2Ki 25.27).There -
was sacking at the temple.Zedekiah,with hostile decisions provoked a return of the Babylonians who will
carry bronze of the temple, incensers and vessels to
Babylon. Gedaliah was made governor. He died tragi -
cally,ten Jews struck him down.   -

Oppositions

For the most part, prophets' criticism ,whose object -
was the temple,had the purpose to extirpate the con
vinction that it signified absolute safety.Nathan was -
the first to set oneself against a specific construction.
When the temple was destroied, the prestige passed
to Jerusalem: in Isaiah one finds that from here ori
ginates Lord's word and that justice will come out -

from Zion.Therefore temple's perspective extends in
cluding a **cosmical vision** .Christians will assume the -
sanctuary as symbol for the Church .The promise made in the Deuteronomy (Dt12.11)is that the temple is
Lord's chosen dwelling for his name. The messianic -
expectation is particularly felt after the exile and in
apocalyptic arguments. The history is put in a new -
frame which is universal[109]. Now are treated various
elements leading to s.John's doctrine.   -

In the recent phase one can also arrange the pro
blem of the relations with Samaritans. At the time of
Ezra ,who helps Nehemiah during the rebuilding of -
the city,other contrasts rise.Samaritans block the re -
contruction of the temple and it will be necessary a -
decree by Darius to continue the works with Zerubba
bel's guide. It's manifest that the rank directing the -
sanctuary wants to protect its own identity deriving
from the choice .One had the dedication's feast in -
515 B.C.; Jesus' meeting with the Samaritan woman
shows an aspect of the hostility between Gerizim and
Mout Zion in real terms. In the ellenistic epoch it will
be formed the party of chasidim (pious people) who -
represent the spiritual heirs of Zadok.

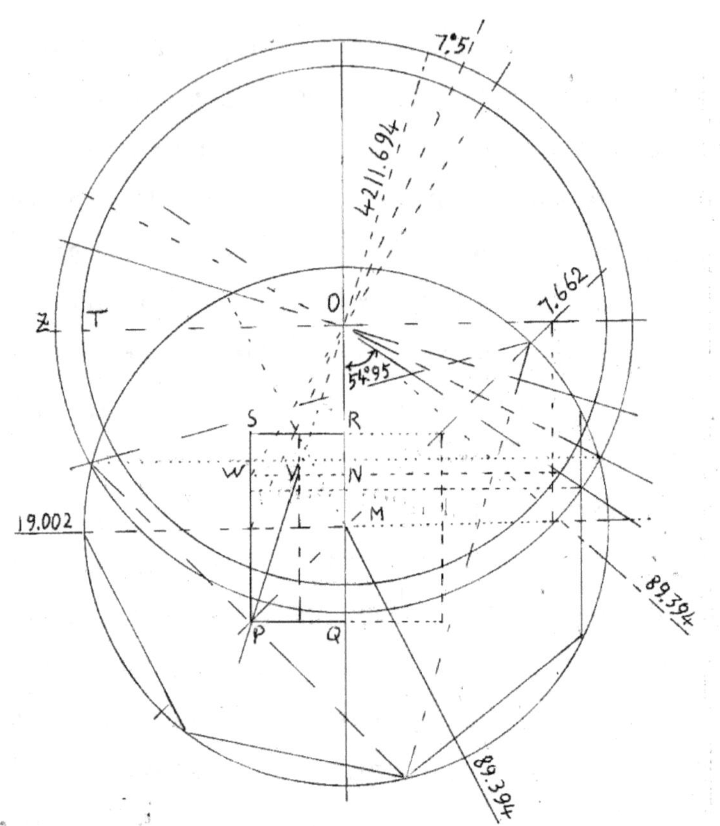

Fig.12PQRS=Ezekiel's(theoretical)sanctuary. The tem
ple of Solomon is the dotted rectangle in the upper
half of PQRS. WN =medial of Solomon's temple(p.54).

$W\hat{V}P=72°$,ZO=77mm., $\dfrac{30.24}{0.63}=\dfrac{360°}{7°.5}$ 89.394+0.63=90.024

[90.024-7.662=82.362] $\qquad \dfrac{30.24}{82.362}=\dfrac{360°}{980°.5}$

Cosmic symbolism

To interpret a possible meaning of the measures in - tervening in temple's construction, according to De Vaux ,one mustn't set aside the historical datum - such as it's understood in Israel's conception.One has most of all to consider both the structure and other - elements elsewhere. Neither one should leave out - the fact that measures for dates were everywhere - performed because of the feasts that had to be cele- brated. This requires angles' knowledge and therefo re it isn't possible to sustain Jews hadn't any docu - mentation from the 8[th]century to Alexander's epoch : actually part of their culture will be found in metri - cal ratios for David's citadel .As example,we explcitily quote a passage from the Midrashim[110]: <The ekal is - at the center of the sanctuary, and the ark is at the ekal's center[111]; the foundation's stone is found befo re the ekal,and is called in this way because the world was founded from it>. It introduces the ekal in classic sense ,as that part (laying before the veil) whose cen- ter in Fig.12 is V, if the line WN is the medial of Solo - mon's temple .   -
After,in the same speech, the Ark is put at the ekal - center ,which means that (Fig.9, p.52) Solomon's ark is in the Most Holy Place at west,and lays there on - the medial of the temple,temple called Ekal as a who

le (p.117).Besides it seems that the Midrashim's text refers to a metric particularity or a natural context - (p.54;p.118) with <the world was founded from it>. -

Temple and Church

Within the ambit of salvation's history,from the temple,spiritualized in the prophetic vision,is looming a - spiritual cult that will manifest itself with the promise of Christ who preaches the reign.In this perspective one finds a synthesis about both the priesthood(considered a holy vocation) and sanctuary, respectively - given by the letter to the Hebrews and the new Jerusalem of heaven. Jesus is the true lamb. He doesn't - need to use animals to get a propitiation and to be - mediator in his Father's presence. -

Owing to the faith one has admission in the reign with Jesus,who acts as a guide to the chosen nation - to build a spiritual construction made of believers, living stones. To the offering 's loaf of the Old Testa - ment is substituted the true bread descended from heaven.The time to wait is projected toward the futu re which will mark Christ's victory on death and evil's darkness. In the Apocalipse recur the prophetic motives and we find the types and signs constituting revelation's leading threads. Convocation of the chosen is called into the celestial Jerusalem where the Lamb irradiates a divine light. God becomes the temple of a city where every earthly tribulations have disappeared. The thriumph of the good,which proceeds from Christ in the reign definitely delivered to his heavenly Father,is exalted. Here,the sacred presence is made clear in the lighting which symbolizes power and holiness.It's Lord's glory pervading the whole, and the city is full of holy splendour.

The ancient pact fixed in Noah's time is found in the sign of a new pact which is the city. With the seal (=the spirit)impressed in the soul one has the likeness of the chosen with that image created by God,and it's realized Christ's complete sovereignty.The choice be comes a name on persons' front and the salvific grace is Lamb's illumination. The risen Christ as propisation has substituted the rite of the Debir. In the prologue of the 4th gospel, the incarnation estabilishes God's dwelling among men.With the apocalyptic revelation, the hope of definitive Lordship reaches the accomplish ment when it's quoted the judgement on the second coming[112].
The expectation sense for what hasn't been seen as yet frequently recurs in the New Testament. The Lord manifests his presence to whom is seeking him.

This hope nourishing people's history catches the -
eye in the events' narration.Jesus' sovereignty at his
second coming marks the coherent development of
a hope which has moulded the spirit and heart of -
Israel. Next to hopes so diversified we find promises'
enrichment[113]: to Abraham is proposed a place rich
in everything , at the Sinai is manifested the gift of
priesthood, during David's reign it's estabilished the
field of kingship.Isaiah will announce the message re
vealing God's presence. The misfortunes are balan -
ced through a great faith in the unique God who assu
re Israelites of his support .   -
That communion which could be partecipated in the
grace of Christ ,manifested in the eucharistic sign,in -
the Apocalipse, becomes a definitive partecipation to
the divine glory.   -

Observations

While we considered the various aspects of the covenant,invitation to holiness,we didn't neglect the ritual
elements that were at sermon's base .The revelation
has always been regarded as a historic redeeming -
process and the mistakes are sistematically outlined,
for example those of the apocalyptic theories. In or -

der to better understand the cult's importance, the - speech about the celestial hierarchy has been remar - kably widened with a continous reference to cheru - bim. They have several features regarding both the just behaviour and soul's multiform wisdom. The cos mic symbolism is carefully treated; convincing ideas on this subject are based on a passage of the Midra - shim (p.89).For the sanctuary's theology it was neces sary to introduce the concepts of choice and cove - nant ,taking the various cultural possibilities into ac - count too. The finitude,as aspect of our existence re- quires an adequate anthropology if the reasoning - hasn't to fall through a philosofy materialistic in cha - racter.No proposals are made for the construction of a humanism;nevertheless the eschatological aspects are appropriate to guide men's expectations through an activity of the individual believer in the earthly ci- ty where the salvation is operating as yet.   -

Conclusion

The phenomenon of the Jewish cult has been widely looked into.The outline of Ezekiel's temple has a defi nite linkage with that of Solomon .It's possible today (2014) superimpose it on a photography of the NASA(American Agency for the outer space) to observe[114] important stellar correlations(p.106).   -

The analysis becomes wider including the dimensions [(11x33)x10] of the exodus' tent (p.52).   -

Covenant is a leitmotif for Jews.In a theology of this pact it must always be taken into account the fu ture coming of the foretold reign.With this idea the re is reference to the origins as one finds in hymns - christological in character. The Reign's programma - tic[115] message included in terms such as justification and grace respectively correspond to choice and bri de image of God's people. All becomes possible ow- ing to Lord's sovereignty, Lord who operates through the history and by his virtue gives rise to new creati on. For the promises relative to the pacts with David and Moses no discordance is found.Neither we must compare Deuteronomy's terms, particularly those of the Decalogue with those of a pagan agreement[116]. - David has to trust in God, but he isn't obliged to him with a vassal's[117] pact to get his benevolence.   - If one thinks to the ancient context in which Jews o perated, one finds the idea of a pristine peace to be reached anew. In this considerations' ambit one has to search the reason of a covenant which is God's - mysterious decree.From Psalms we deduce[118] that - Hannukah's feast was a celebration of the pact.In o- ther terms God's sovereignty was testified in a cultic form. The peace is a gift of Yahweh's munificence -

and has nothing to do with a treaty.Covenant's conditions are exhortations. Their content is the follow
ing: one has to do good[119] and to avoid evil as well. -
Jeremiah shows the roots of Yahweh's spiritual cult.
After the downfall with the exile the Jews must place
hope in the Lord,able to transform the hearts.   -
It appears a spiritual eschatology still absent in Job's
book. That God who created the firmament, according to the Deuteronomistic tradition, won't repel his
chosen. Hence an ideal model is transmitted :that of
the new creature.To take into account a temple's cos
mic symbolism means to outline an aspect of God's -
action,object of the revelation, which isn't exclusively
theoretical or descriptive. It has redeeming value.   -
Fixing at the same time, an individual, collective and
eschatological horizon from world's origins implies a
values'ordering which,respectful of the sacred, exclu
des[120] every gnostic idea and docetism.Moreover it -
overcomes the myths, some known heresies and it's
avoided the possibility to have both Barth's interpre
tation and that of R.Bultmann:christology,pneumato-
logy and eschata form the notes of an unique divine
message speaking of man's integral vocation[121].Saints
call our attention to look at the human homeland of -
Heaven; suggest one should also look to the Church -
for help, because our Lord has a dwelling and sacra-

ments[122]in it. Peace is present and operates as yet in the world ,in the symbolic millennium in which the - Church reigns with Christ.Again we find the germs of s.John's enuntiations of the tension between what is available at present and the future that has to come; tension synthetizing Christ's mission[123].
Adopting a common terminology,christianity allows - to deduce a sound anthropological doctrine, outlined with speechs on redemption and grace,which take for ce from the sermon of the Mount.   -

This is the parallel between Jewish temple and glorified Christ :Jesus after the dialogue with Nicode- mus represents Jacob's new ladder, able to unify - Heaven and Earth. In the first creation and later one can remove every pessimism or presumed separati- on between God's city and earthly city,finding the joy to advance with a construction of a world which isn't in contrast with God,but has been given to men up to Jesus' second coming .   -

Appendix A

Moon's phase

Earth and moon,going along their orbits ,undergo a -
change in a day respectively given by
$(360°/365.2422 \text{ days})=0°.985647332=b$
$(360°/27.3216 \text{ days})=13°.1764=a \Rightarrow (a - b) =$
$12°. 19075267$ a day.   This difference (a-b) will be
found periodically again after
$360°/12°.19075267=$**29.53058025** days;   whence
$12(29.53058025)=$**354.366963** .

PRACTICE:On both*December 10 1996 and Gennuary
17 1999 the dark side of the moon faces the earth.
It's **<u>new moon</u>**. The days in this interval are given by:
21 days during December 1996
2(365.2422)days for the years 1997 and 1998
17 days of Gennuary 1999
As a total amount one has **<u>769.4844</u>** days
Let's find the date Gennuary 6 2.000 as follows :
17 days (of Gennuary 1999)+354=371;then 371-365=
=Gennuary 6 2000 .
PHASE(=position) expressed by degrees :
769.4844+354.366963=1123.851363 days .Phase=
=(1123.851...)(12°.19...)=13699°.74369≅360°(38)+
+19°.74 .⇒Phase:
 *Salvo De Meis Jean Meeus- Almanacco
Astronomico 1996-hoepli Ed.-Milano(Italy),p.85 .

Simpler procedure

1)Dropping 360°(38) cycles, the result is 19°.74

2) If the angle of 19°.7 is made by the line drawn -
from the center of a cirle having a radius of 85mm
on the x axis ,the point P on its terminal side is speci
cified by giving its cartesians coordinates:
 (79.6; 29.8mm).(H.Weyl's formula:)

3)Phase=$\dfrac{79.6}{\sqrt{79.6^2+29.8^2}} \cong \dfrac{79.6}{84.995}$=0.9365 ;$\dfrac{79.6}{85}$=0.93647

[Phase=cos19°.74=0.941] .

Exercise: Find moon's phase on September 23 2010.
Solution:We have to consider 21 days of December
1996; adding 13(365.2422) days one reachs the end
of 2009.In fact (2009-1996)=13.The span from
Gennuary 1st 2010 to the September 23 measures
266days. Moreover 2000-2004-2008 are leap years,
each with 366 days; ⇒one has 3 days to add to the
previous: 266+21+3=290

13(365.2422)=**4748**.1486

⇒4748.1486+290=**5038.1486**

5038.1486 − **769.4844 0=4268.6642 (**p.97**)**

4268.6642(12°.19075267)=52038.22949=α

In the difference , above,**769.4844(=new Moon)*** is

*Observers'Handbook 1995-Editor Roy L.Bishop-The
Royal Astronomical Society of Canada -136 Dupont
Street-Toronto-Ontario M5R1V2 .

considered zero.   α/360)=360°(144)+198°.2

[Phase=cos198°.2=0.94997205; or :]

Simple evaluation(198.2=180+**18**.2)

Consider a circle with radius of 85mm.We put the a -
gle 18°.2 made by the line originating from its center
on the x axis;such a line intersects the circumferen
ce at the point P(80.5;26.8). Using H.Weyl's formula,

Phase $= \dfrac{80.5}{84.995} = 0.9471 \cong 0.95 (\equiv$ **full Moon**$)$.

Ancient calendar of Dionigi(il Piccolo) It dispels every
doubt about December 25.

Example: Find the date of Easter 2011.

Dionigi fixes[124] the date December 25 as 0(= year ze -
ro,birth of Jesus Christ) on the number 753 .

To obtain Easter Sunday one has to find March 21.

The days from March 21 to December 25 are 279;hen
ce (753-279)=**474** represents March 21.

This same date will occur after 365.2422 days.

   First of all, we want to have the Easter 2010(known
as yet):in such a way one identifies a Sunday.

PRACTICE:

1) 2009years=2009(365.2422)= 733771.5798 ;
2) 2000 is a leap year(with 366 days)

3) $\dfrac{2009}{4}$=502.25=number of leap years .

4) [**474+365.2422(**=March 21 of the 1styear)]+

+733771.5798+502=**735112.822**=March 21 2010 .

5)$\Rightarrow$(Known Easter 2010)=

=735112.822+14=**Sunday**=735126.822

6)[(March 21 2010=)735112.822]+365.2422=

   March 21 2011   =735478.0642= [(**a**)]

7) **Definition**:   **Easter** will be[125] < the first Sunday after the new Moon after the fourteenth day after March 21st >.

      Hence,starting from **769.4844(new Moon)** with n(**29.53058025**) where n is integer, one must little exceed [(a)]+14=735492.0642 .

8)769.4844+n(29.53058025)=735483.3607485

9)Adding another new Moon,that is **29.53058025**

 one finds 735512.89132875=[(b)]

10)The elapsed weeks since Easter 2010 are 55 :

[(b)]-735126.822$\cong$386.069$\cong$7(55.1)

7(55)=385; 385+735126.822=Easter

2011=735511.822

11)735511.822-735478.064=33.75 $\approx$34 days .

....

Other procedure

The gap between September 23 to   March 21 2010
is made of 186days.

5038.1486-186=4852.1486=March 21 2010 (**p.98**)-
Known Easter 2010=4852.1486+14=**4866.1486**

March 21 2011=March 21 2010+365.2422=

=**5217.3908**;   **new Moon** =

 =769.4844+n(29.53058025)=5228.60201795

2000 2004 2008 yield an enhancement of 3 days

5231.602 ;5231.6020+14=5245.6020 .

(5245.6020 - 4866.1486=379.4534≅7(54.2)≅54

weeks;

For a week later, according to the definition(p.100),
55weeks=385 days;whence

4866.1486+ 385=5251.1486=Sunday .

5251.1486-**5217.3908**=33.7578≅34 days .

Thus, **April 24 2011** is Easter day .

How to controll the evaluation

(Butcher's method of the ecclesiastic calendar[125]):the reader who wanted to obtain other Easter days with -Dionigi's procedure should make a comparison with the following one .Only the knowledge of the year is -needed.

Example:Easter 2011 . $\dfrac{2011}{19}$ =105.8421053 ;hence

	Integer part	remainder
	105	a=19(0.8421053)
$\dfrac{2011}{100} = 20.11$	b=20	c=100(0.11)=11
$\dfrac{b}{4}$=5.000	d=5	e=4(0.000)=0
$\dfrac{b+2}{25}$=1.12	f=1	
$\dfrac{b-f+1}{3}=\dfrac{20}{3}$=6.6...	g=6	
$\dfrac{19a+b-d-g+15}{30}$=10.9333333		h=30(0.9333333)=28
$\dfrac{c}{4}$ =2.75	i=2	k=4(0.75)=3
$\dfrac{32+2e+2i-h-k}{7}$=0.714285714		L=7(0.714...)=5
$\dfrac{a+11h+22L}{451}$=0.96.....	m=0	
$\dfrac{h+L-7m+114}{31}$=4.741935484		p=31(0.74193584)=23

In this case,the integer part 4 means April;p+1=23+1= 24= Easter day.Whenever one finds 3 instead of 4,the month is March.

Note:For Metone cycle[126] (5th century B.C.)
 (29.53 days)(235)=6939.55 ;365.24(**19**)=6939.56 ,
and therefore the phase 29.53 returns at the same
value in the same day .In fact
29.53058025(235)=6939.686359
365.2422(19)+4(=leap years)=
=6943.6018=7(**991.9431143**).

To study calendars in depth ,it's useful to know that
14 types[127]of gregorian calendars follow one another.
In the first of them Gennuary begins with Monday.
By a regular ordering ,only with the calendar number
eight Sunday appears again. Moreover up to number
7 they contain a February with 28 days; for the re -
maining calendars one must add another day to Fe -
bruary .
Some results are given by:
2014 4 ;2015 5 ; 2016 13;   2017 1 ;
2018 2 ;2019 3 ; 2020 11;   2021   6 ;
2022 7 ;2023 1 ; 2024 9 ;   2025 4 ;
2026 5 ; 2027 6 ; 2028

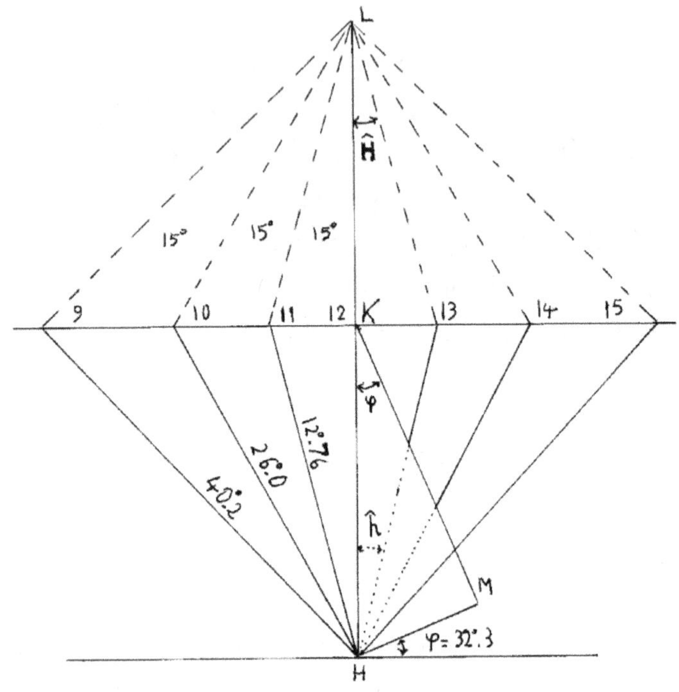

Fig.13-Vertical sundial at Gibeon.
LK=MK; $\tan\hat{h}=\cos\varphi\tan\hat{H}$.
φ=32°.3

$\hat{h}$=angle read on the sundial owing to sun's motion.

Vertical sundial[128] at Gibeon(cosφ=0.845261833)

Hours			$\widehat{H}$	$\tan\widehat{h}$	$\widehat{h}$=read angle
11 and	13		15°	0.226487225	12°.76
10	"	14	30°	0.488012146	26°.01
9	"	15	45°	0.845261833	40°.206
8	"	16	60°	1.46403644	55°.665
7	"	17	75°	3.154560106	72°.41
6	"	18	90°		90°
5	"	19	105°	- 3.154560106	-72°.41
4	"	20	120°		-55°.67

Analogously ,in a horizontal sundial one finds

$\tan\widehat{h'}$=sinφ$\tan\widehat{H}$;sinφ=0.534352349

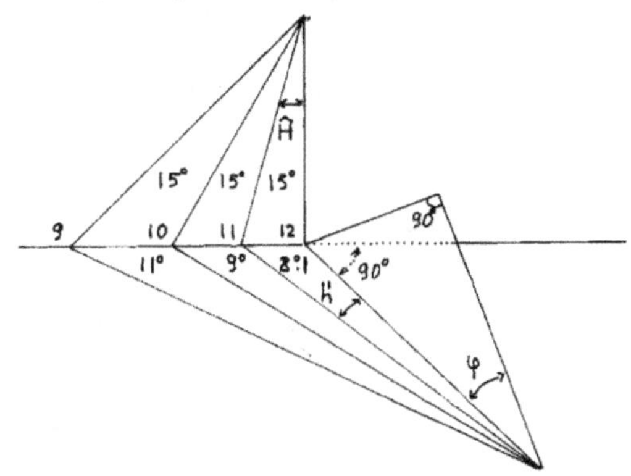

Fig.14-Horizontal sundial at Gibeon.

$\widehat{H}$	$\tan\widehat{h'}$	$\widehat{h'}$
15°	0.14317929	8°.1
30°	0.308508472	17°.1
45°	0.534352349	28°.1

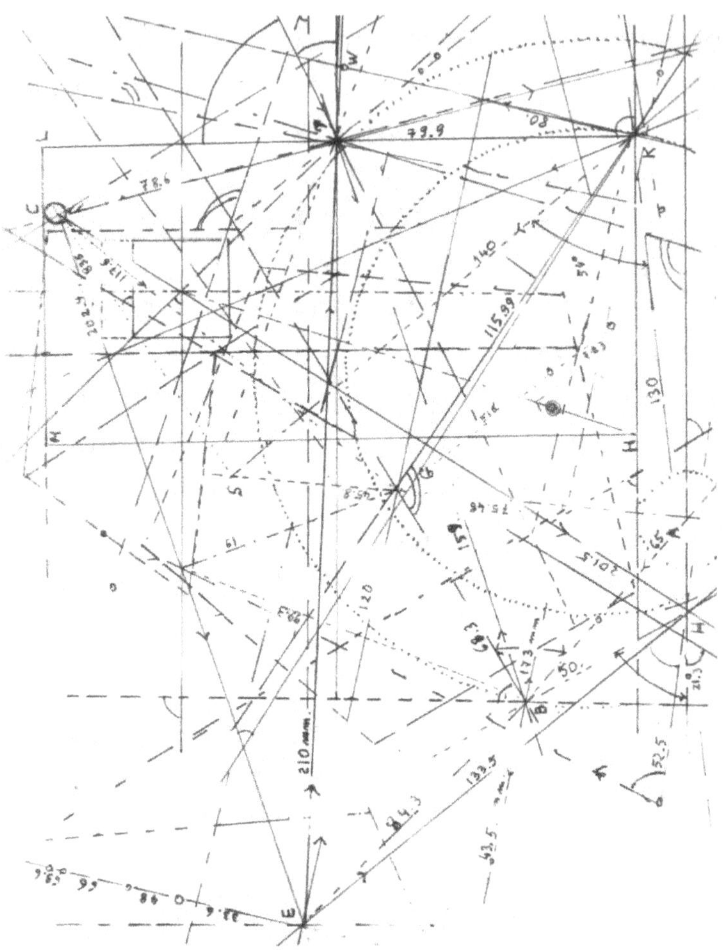

Fig.15-HKLM=Ezekiel's temple in a domain containing some Black Holes[Adapted from SPAZIO Hobby&Work[129] Publishing n.23(2013)p.272-273-Milano-Italy]. Rees-Black Holes in Galactic Centers –Scientific American[=S.A.]Nov1990. King-Globular Clusters [S.A.]June1985.Foster Nightingale-A Short Course in General Relativi ty-Springer Verlag-New York 1995.
[Distances=mm. $99.6459e^{\pm k0.0368}$]

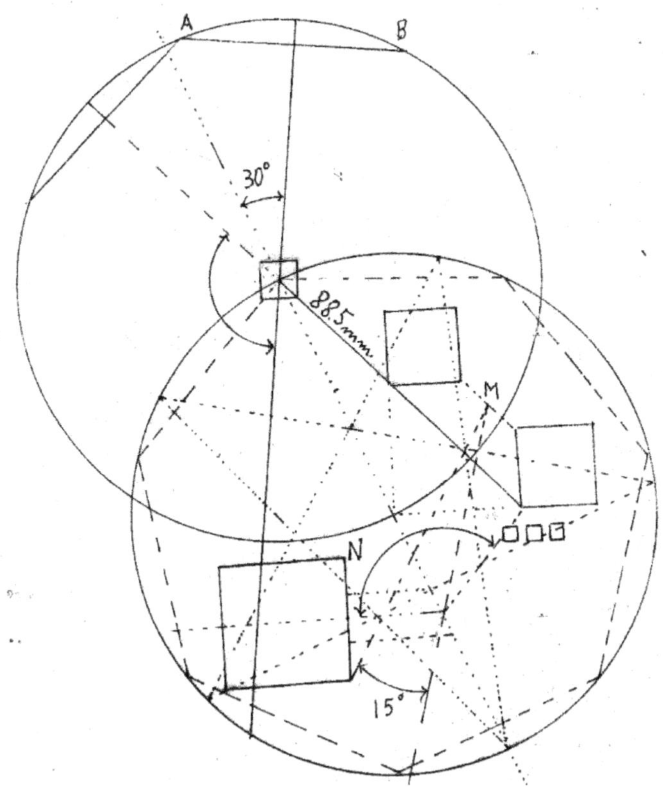

Fig.16-Egyptian pyramids(4 squares) seen from ab<u>o</u>ve.

Sir Fred Hoyle in his book<Stonhenge>reveals[130] that in ancient times they used a circle with **56** peripheral holes to deduce eclipses both of moon and sun.

MN=**56** cm.- An hour has elapsed with 15° to find the **Sphinx**.

2184=**56(39)**=168(13) .

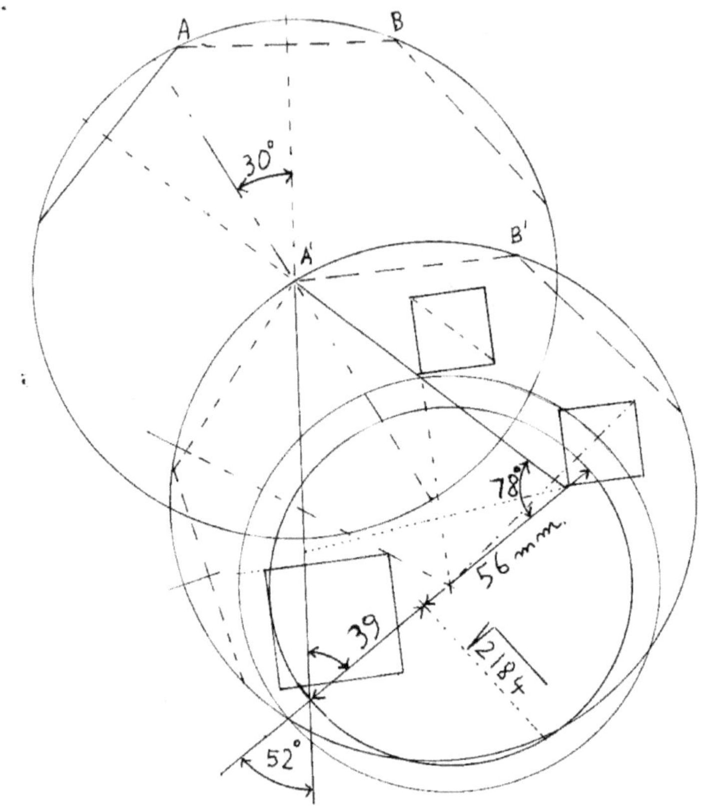

Fig.16b-Some details. (p.107)

2184=8(**273**)=6(**364**)=6[7(52)]days .

364=**52weeks**

[With **27.3216** days the moon completes the motion along ist own trajectory(G.Romano,p.83).]

Fig.17-Part of an Egyptian city.
(Brooklyn Museum-New York).

Egyptian table[131]

 J.Gillings guesses that Egyptians were possibly lear -
ning numbers' combinations by the use of the follow
ing table:

2911	2810			**246**	**235**
3912	3811			347	
4913	4812	4711	4610		
5914	5813	5712	5611		
6915	6814	6713			
7916	7815				
8916					

Actually,it's a code that we now decipher:on the first column(six steps)(p.109) $\Rightarrow$ 6(1001)= 6(11)(91);hence 4(91)=364.

Table: 234+n11 , where n is integer.
[235.52(100)=126.32+109.20]

234	311	388	465	542	619	696	773	850
245	*322*	*399*	*476*	**553**	*630*	*707*	*784*	*861*
256	333	410	487	564	641	718	795	872
267	344	421	498	575	652	729	806	883
278	355	**432**	*509*	*586*	663	740	817	**894**
289	366	443	520	597	674	751	828	
300	377	454	531	608	685	762	839	

$$\frac{30.24}{1.092}=\frac{360°}{13°}$$

235.52+10.92=**246**.44 (p.109)

432+4(30.24)=**552.96**   (p.37)

432(7)=3024

[13.10+5.46(14)=89.54 (p.118)]

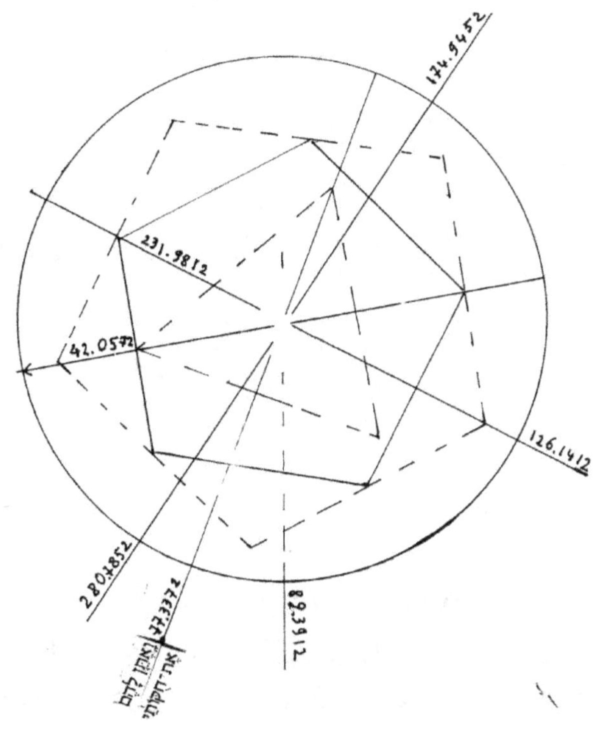

Scale : Along the circle one adds 7(30.24)= 211.68 after a rotation through 360°.

Fig.18-Structural elements in Ez 20 .

From the sequence (Ez20.6); (Ez20.11); (Ez20.18); (Ez20.25); (Ez20.33); (Ez20.40) $\Rightarrow$

6x7=42 11x7=77 18x7=126 25x7=175 33x7=231;

40x7=280(p.18). $\left\{ \begin{matrix} Hebrew[Ez\ 20.11] = \\ statutes \end{matrix} \right\}$ put at 11x7.

$126.1412 + \dfrac{211.68}{4} = 231.9812$

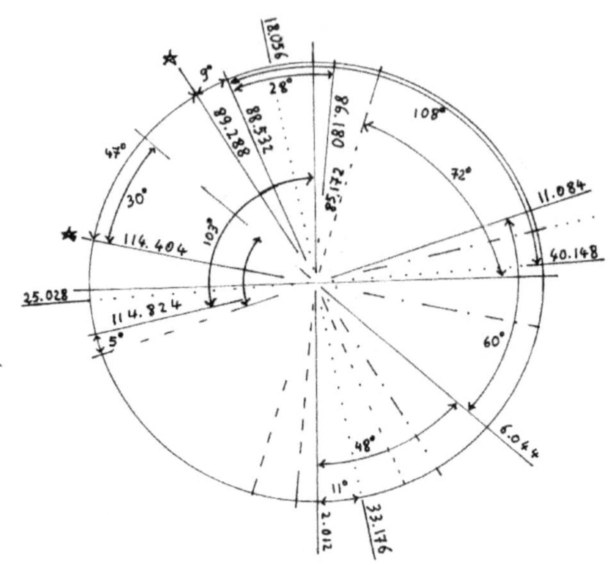

Scale: 2°⇔ 0.168;   360°⇔30.24

Fig.19- Structure of (Ez20) (= salvation's classic histo -
ry). Starting from 40.148, after an angle through 108°
one finds 88.53 fundamental reference point for eve
ry calendar.$\sqrt{110^2 + 33^2} = 114.8433716$ =d
(diagonal of exodus' tent).

4201.614-d=91(45)(**0.997990385**)

d-7.662=107.1813716 ; $\dfrac{30.24}{107.813716}=\dfrac{360°}{1275°.96871}$

$$\left\{ \begin{array}{c} 28° \equiv 14(0.168) = 14[7(24)]/1000 \\ 30.24 = 7(4.32) \Rightarrow 7(30.24) = 211.68 \\ \textit{the circle has to be often subdivided in 28 parts} \end{array} \right\}$$

(p.22;p107)

112

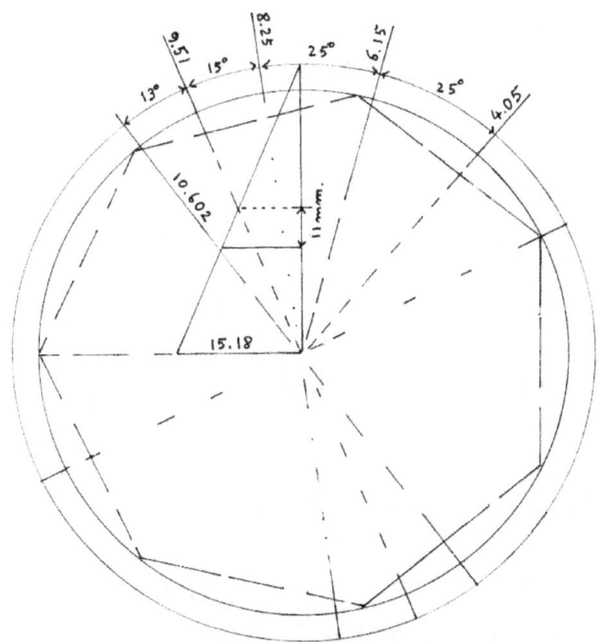

Fig.20-One gets 13°⇔1.092 for the Uruk's ziggurat[132] which is a truncated pyramid. Half of it is shown here.

In fact (10.602-9.51)=1.092 ; $\dfrac{30.24}{1.092}=\dfrac{360°}{13°}$.

$$30.24\,|\,360°$$
$$2.73\,|\,32°.5$$
$$2.52\,|\,30°$$

$$24(\mathbf{365.2422})-\mathbf{34.02}\,|\,\alpha=360°(288.7497619)(\mathbf{p.38})$$

$$\left(\dfrac{30,24}{24(365.2422)-34.02}\right)=\dfrac{30.24}{8731.7928}=\dfrac{360°}{\alpha}$$

$$(10.602-4.05)=6.552 \Rightarrow \dfrac{30.24}{6.552}=\dfrac{360°}{78°} \quad (78°\text{ at p.37 too})\,.$$

Solution* for the connections of p.106

Let's find the eigenvalues λ through the equation

$$\begin{vmatrix} -\lambda & 0 & -1 & -1 \\ 2z & -\lambda & -1 & 1 \\ 1 & 1 & -\lambda & 0 \\ 1 & -1 & 2z & -\lambda \end{vmatrix} = 0 \qquad \text{which means}$$

$$(-\lambda)\begin{bmatrix} -\lambda & -1 & 1 \\ 1 & -\lambda & 0 \\ -1 & 2z & -\lambda \end{bmatrix} - (0)\begin{bmatrix} 2z & -1 & 1 \\ 1 & -\lambda & 0 \\ 1 & 2z & -\lambda \end{bmatrix} +$$

$$(-1)\begin{bmatrix} 2z & -\lambda & 1 \\ 1 & 1 & 0 \\ 1 & -1 & -\lambda \end{bmatrix} - (-1)\begin{bmatrix} 2z & -\lambda & -1 \\ 1 & 1 & -\lambda \\ 1 & -1 & 2z \end{bmatrix} =$$

$$-\lambda[\,-\lambda(\lambda^2) + (-\lambda) + (2z - \lambda)\,] +$$

$$-[\,2z(-\lambda) + \lambda(-\lambda) - 2\,]$$

$$+ [\,2z(2z - \lambda) + \lambda(2z + \lambda) - (-2)\,] =$$

$$= \lambda^4 + 4\lambda^2 + 4(1 + z^2) = 0$$

Hence $\quad \lambda^2 = -2 \pm \sqrt{4 - 4(1 + z^2)} = -2 \pm 2iz = 2(-1 \pm iz)$. After a change of scale, we put z=1 and the solution will be

$$(\lambda')^2 = -1 \pm iz = \varrho e^{\pm i\varphi} \quad \text{where } \varrho = \sqrt{2},\, \tan\varphi = 1 \Rightarrow$$

$\left|\dfrac{\varphi}{2}\right| = 22°.5$ .With only two values of λ', by adding,

one has $\sqrt[4]{2}(2\cos 22°.5\,) = 2.197368227 = T =$
=Trace on a plane . With multiples nT one obtains significant levels for the Fig.15 of p.106 concerning a domain with some Black Holes.

* V.I.Morrone-Matrices for Black Holes-
https:/www.youcanprint.it

Meaningful levels in fig.15(p.106)

1)Distances **from G** :
12.1T=26.5881mm.[(Andromeda) galaxy not
represented at p.111]
20.85T=45.815mm.(galaxy S)
34.35T=75.479(galaxy A)
27.75T=60.97   ,52.75T=115.911,   53.5T=117.55,
54.75T=120.30,   23.5T=51.63

2)Distances **from B**:
22.75 T=49.990;   72.5T=159.309 ;
78.75T=173.04 mm.

3)Distances **from K**:
36.5T=80.20 , 59.25T=130.19,   63.75T=140.08 mm,
36.35T=79.87 ,   36.5T=80.20 ,
59.25T=130.19 ,   63.75T=140.08

True noon [133] **in the vertical sundial**

For a sundial,a fictious sun moves at constant speed along the equator.To get the true time V it's necessary a correction E, variable during the year,that can be obtained with time's equation.Moreover,one must take into account the local meridian , different from - that of the zone.

Each zone extends 7°.5 on either side of its central meridian.

Example:

1)Gibeon's longitude is 34°.6885;but the zone's meridian is at 37°.5.
Whence(37°.5-34°.6885)=2°.8115=M .
2)The equation of the time[136],if the correction is made on December 23,yields E=-1.1 minutes=-1.1/60= =-0°.18
3)V=M-E=2°.815+0°.18=3°
4)The correction is of 4 minutes per degree;hence it is 4x3=12minutes
5) $\dfrac{60minutes}{12minutes} = \dfrac{15°}{3°}$ ⇒ noon's notch must be put at 3° as respect to the vertical .

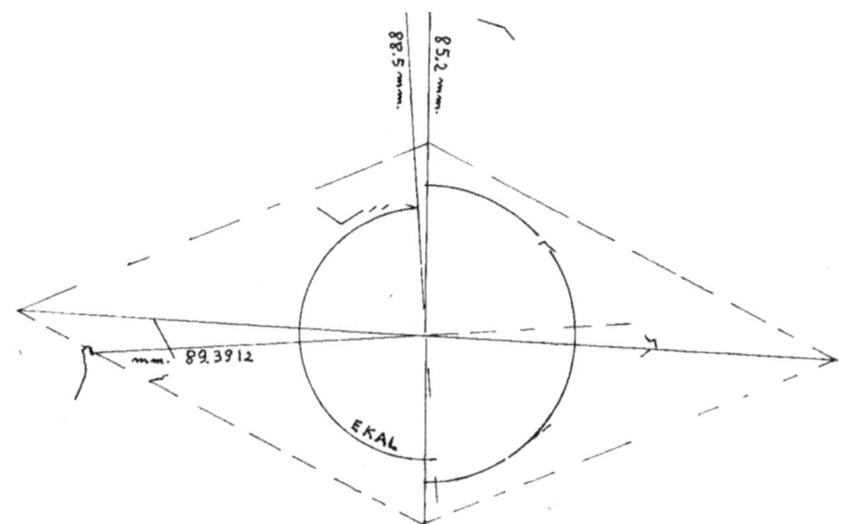

Fig.21-Astronomic conjuncture[134] at Jerusalem.
16440=7(**2348**.571429) (p.24) ,   2184(31)=**67704**

$\Rightarrow$ 16440+67704=84144 ,[$\frac{2184}{4}$=546]

84144+546(4)=**852**36;**852**36+546(6)=**885**12

16440-9.264=16430.736 , $\frac{30.24}{16430.736}=\frac{360°}{195604°}$

[1051,8-851.6=2(100.1)=2(91)(1.1); $\frac{30.24}{3(20.02)}\frac{360°}{715°}$]

The lowest vertex is a corner of the citadel; the rea -
der could identify more points of the area with a pen-
tagon centered as the rombus.
Ekal =**whole palace***and also part of it.
*R.K.Harrison- Biblical Hebrew-NTC Publishing Group-
4255 West Touhy Avenue – Lincolnwood Chliffs -
(Chicago) 1993,p.150.

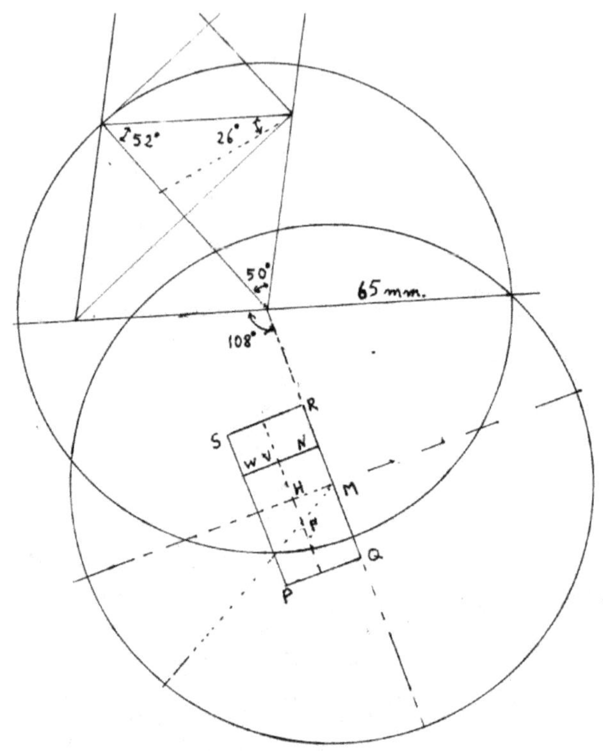

Fig.22-Moon at the temple.Ezekiel's temple PQRS.
Note VP and analogous line in Fig.15. $W\hat{V}P=72°$.

$VM=mm.11\sqrt{2}$, $VP=\sqrt{11^2+33^2}=34.78505426$ (p.112)

89.3912-VP=54.60614574≅54.6 ; $\dfrac{2.73}{54.6}=\dfrac{32°.5}{650°}$ (p.117).

A flight 13.10+5.46(4)=34.94 ≅34.78 ,from Kaabah's roof to V;
actually m.13,10 is the hight* of this building.

[MH=11=MFcos29°.45;$\dfrac{MF}{MV}$=0.81203=λ ,$70e^{-0.09\lambda}$=65.06]

*Cesar Abdel Salomon-Islam e Musulmani?-Cerebro Editore.

Appendix B

Notes in the text:

1-E.Cassirer- Simbolo, Mito e Cultura – Laterza Ed.-Roma 1985,p.151.

2-C.Skalicky- Alle Prese con il Sacro- Herder Ed.-Roma 1982,p.210.

3-F.Barbiero- Alla Ricerca dell' Arca dell' Alleanza -SugarCo Ed.-Milano 1985,p.169.

4-D.G.Bressan-Samuele-Marietti Ed.-Torino 1960,p44

5-A.Mistrogiro- L'Arte Sacra- Messaggero Ed.- Padova 1983 .

6-G.Damasceno - Difesa delle Immagini sacre - Città Nuova Ed.-Roma 1983,p.63 .

7-G.Damasceno,p.44 .

8-A.Lancellotti et al.- La Distruzione di Gerusalemme del 70- Collectio Assisiensis , 8 ,Studio Teologico <Por ziuncola>Assisi 1971,37 .

9-M.Grant- L'Antica Civiltà d' Israele – Bompiani Ed.-Milano 1984 .

10-Y.Aharoni M.Avi Yonah- Atlante della Bibbia -PIEMME Ed.-Casal Monferrato 1987,p.48 .

11-H.R.Mills - Practical Astronomy - Albion Publishing 1944,p.110 .

12-S.De Meis J.Meus- Almanacco Astronomico 1955-Hoepli Ed. Milano .13- Holy Bible with Apocriphal/Deuterocanonical - Books- New Revised

Standard Version-American Bible Society-New York 1989,p.200 .

14-Supplements to Vetus Testamentum,VII p.8 –J.Brill -Leiden.(See also:Bernhard W.Anderson- Understan - ding The Old Testament-Fourth Edition;Prentice-Hall, Inc.- Englewood Cliffs,New Jersey 07632,1986,p.224)

15-G.Battista Proja- Battistero Lateranense-Tipografia Poliglotta Vaticana 1990 .

16-D.G.Bressan – Samuele - Marietti Ed.-Torino 1960, p.110 .

17-F.Brown et al.-Hebrew and English Lexicon of The Old Testament-Oxford University Press-London 1962, p.847 .

18-C.Martini-Chiesa– Vescovo – Martirio-Paoline Ed. - Roma 1983,p.23

19-Scientific American,July 1985,p.82 .

20-D.G.Bressan – Samuele - Marietti Ed.-Torino 1960, p.57

21-A.Issar-Fossil waters under the Sinai peninsula.see Scientific American July 1985,p.83 .

22-Encyclop. Amer. 21 – American Corporation -New York 1977,p.198 .

23-I.Steinhorn et al.- il Mar Morto: see < Le Scienze > (=translated Scientific american)Dicembre 1983,p.96 .

24-J.Perrot – Siria – Palestina , 1, Nagel-Ginevra 1977, p.108 .

25-<Le Scienze>(=translated Scientific American) Gen
naio 1984-Milano-Italy,p.68

26-G.Romano- Introduzione All' Astronomia - Franco
Muzzio Ed.-Padova 1985 .

27-H.Robert Mills-Practical Astronomy;A User-Frien -
dly Handbook For Sky Watchers - Albion Publishing -
Chichester 1994,p.110 .

W.Schroeder-Astronomia Pratica– Ed.Longanesi & C.-
Milano 1979.

28-J.Oates – Babilonia - Newton Compton Ed. - Roma
1984 , p.230 .

29-A.Parrot-Archeologia della Bibbia-
Newton Compton Ed.-Roma 1978,p.23 .

30-A.K. Kempiski - Siria - Palestina, 2 , Nagel - Ginevra
1977,p.74 .

31-E.R.Galbiati et al.-Atlante Storico della Bibbia e del
l'Antico Oriente-Jaca Book-Roma 1983,p.72s .

32-R.Harker- il Mondo della Bibbia -Newton Compton
Ed.-Roma 1981,p.41 .

33-N.Borrelli – Tradizioni Aurunche-Centro Studi Min
turnae-Perugia 1984,p.98 .

34-Cotterel-Civiltà antiche-Editori Riuniti-Roma 1981,
p.150 .

35-A.Paul -Photo Guide de l'Ancien Testament-Paul -
ton 1976,p.56 .

36-J.B.Metzelershe-Real Encyclop. 1927,p.102 .

37-M.Haran- Tempels and Temple Service in Ancient Israel-Clarendon Press-Oxford 1978 .

38-J.Mauchline1 and 2 Samuel-Oliphants Ed.- London 1971,p.50 .

39-P.Kyle-Mac Carter Jr.-Samuele-p.74

40-R.W.Klein- 1 Samuel -World Book Publisher -Waco Texas 1983,p.4 .

41-The Interpreter's Dic. Of The Bible-Supplementary Volume-Abingdon Press-Nashville 1976,p.329 .

42-Enciclopedia Cattolica,Sansoni Ed.,Firenze-p.589 .

43-J.Emerton - Studies in the Historical Books of the Old Testament-J.Brill-Leiden 1979,p.92 .

44-R.E.Brown- Comentario Biblico San Geronimo-1°, Edic.Cristiandad-Madrid 1971,p.449 .

45-R.K.Harrison - Introduction to the Old Testament - Tyndall Press-p.449 .

46-Kyle-Samuele,p.123 .

47-Kyle,p.58 .

48-H.M.Barstad - The Religions' Polemics of Amos - J.Brill - Leiden 1984,p.199 .

49-R.W.Klein- 1 Samuel-World Book Publisher-Waco Texas 1983,p.143 .

50-G.von Rad-Teologia dell'Antico Testamento-Paideia Ed.-Brescia 1972.p.278 .

51-E.Pace-Dizionario dl Sociologia e Antropologia Culturale-Cittadella Editrice-Assisi 1984,p.334 .

52-J.Parrot - Siria Palestina , 1 ,Nagel – Ginevra 1977 , p.147 .

53-V.Malka – Israele - Edizioni Futuro - Verona 1982, p.122 .

54-C.M.Guzzetti- il Messaggio di Allah-LDC Ed.-Torino 1979.p.12

55-Atti XXVI Settimana Biblica-Gerusalemme-Paideia Ed.-Brescia 1982 .

56-E.Beaucamp-La Bible et Le Sens Religieux de l'Uni - vers-Du Cerf Ed.Parigi 1959,p.33 .

57-R.Harker-il Mondo della Bibbia - Newton Compton Ed.-Roma 1981,p.107 .

58-R.De Vaux-Ancient Israel- Darton Longman & Todd Ed.-Londra 1962 .

59-P.Reymond - L'Eau, Sa Vie et Sa Signification dans l'Ancien Testament-J.Brill Ed.-Leiden 1952,p.235 .

60-J.Uris et al.-Jerusalem- Le Cantique des Cantiques-Doubleday-New York 1981,p.71 .

61- T.Ballarini – Salmi – Dehoniane Ed.-Naples 1978 - p.286

62- Y.Aharoni M. Avi-Yonah - Atlante della Bibbia - PIEMME Ed.-Casal Monferrato(AL) 1987,p.48.

63-R.Bultmann-Teologia del Nuovo Testamento-Que- riniana-Brescia 1985.

64-NIV bible Commentary,Vol,1°,Old Testament-Zon -

dervan Publishing House – Grand Rapids – Michigan, p.1350.

65-A.Dupont- The Essen Writings from Qumran - The World Publishing Comp.-Cleveland 1962-

66-A.Gelin- Les Idées Maitresses de l'Ancien Testa - ment -Du Cerf-Paris 1959,p.52.

67-Peaches Commentary on the Bible-Nelson-London 1962,p.587.

68-J.Dheilly- Dictionnaire Biblique – Desclée - Tournai 1964,p.1158 .

69-E.Beaucamp-La Bible et le Sens Religieux de l'Uni - vers-Du Cerf –Parigi 1959,p.33 .

70-J.Oates-Babilonia,p.107 .

71-Supplements to Vetus Testamentum- Volume du Congress-J.Brill-Leiden,p.48 .

72-Y.Congar – Le Mistère du Temple - Du Cerf - Parigi 1958, p.68 .

73-Eusebio-La Preparation èvangèlique-Du Cerf-Parigi 1975,p.159 .

74-J.Coppens et al.- Sacra Pagina-Congressus Interna- tionalis Chatolici De Re Biblica-Duarlot Ed.- Gembloux 1959,p.536 .

75-F.Pierini- Le Religioni dell'Antichità: see Guida alle Religioni-Paoline Ed.-Roma 1985,p.51 .

76-T.Reinach – Antiquités Iudaiques – Leroux - Parigi 1912.

77-Giuseppe Flavio-Antichità Giudaiche.

78-Kempiski-Siria,67 .

79-M.Strassfeld - The Jewish Holidays - Harper & Row Publ.-New York 1985,p.162 .

80-J.Gutman- The Temple of Solomon -Scholar Press-Missoula 1976,p.705 .

81-D.Gottlieb-il Giudaismo:see <Guida alle Religioni>-Paoline Ed.-Roma 1985,p.239s .

82-A.Parrot- Le Temple de Jerusalem-Niestlé-Neuchatel .

83-J.Gutman – The Temple of Solomon-Scholar Press-Missoula .

84-W.Corswant - Dic. d'Archeologie Bibl. - Delachaux-

85-P.Vanderberg-La Maledizione dei Faraoni-SugarCo Ed.-Milano 1985,p.260.

86-Richard J.Jillings-Mathematics in the Time of Pharaohs-Dover Publications 1982- Originally published at the M.I.T. Press-Cambridge.

87- Pubblicazione Settimanale <SPAZIO> - Hobby & Work Publishing-Milano-Italy;n.23(2013),p.272-273 .

88-De Vaux-Le Istituzioni...,p.284

89-De Vaux-Le Istituzioni...,p.285

90-Parrot-Archeologia della Bibbia -Newton Compton Ed.-Roma 1978,p.83 .

91-Origene-Omelie sul Levitico-Città Nuova Ed.-Roma 1985,p.204 .

92-Oates-Babilonia,p.143 .
93-De Vaux-Le Istituzioni…,p.285 .
94-G.Giannini –Ateismo – Speranza -Città Nuova Ed. –
Roma 1973,p.18 ,
95-B.Gherardini - La Chiesa - Oggi e Sempre -Ares Ed.-
1974,p.34 .
96-Origene-Omelie sul Levitico,p.225 .
97-I.Sanna - L'Uomo - Via Fondamentale della Chiesa-
Dehoniane-Napoli 1985,p.230 .
98-B.Gherardini-La Chiesa…,p.33 .
99-D.Mollat-Principi d'Interpretazione dell'Apocalisse
(see: L'Apocalisse-Studi Biblici Pastorali 2-Paideia Ed.-
Brescia 1967,p.9) .
100-Origene-Omelie…,p.273 .
101-Origene,p.224.
103-Temple=33x11;Exodus'tent 33x110 .
104-De Vaux-Le Istituzioni…,p.314 .
105-Ps.Dionigi L'Areopagita-Gerarchia Celeste-Teolo -
gia Mistica-Lettere-Città Nuova Ed. Roma1986, p.54 .
106-De Vaux-Le Istituzioni ….,p.298 .
107-Ps Dionigi-Gerarchia Celeste…,p.48 .
108-W.H.Schmidt - Dizionario Biblico - Jaca Book Ed. -
Milano(Italy).
109-G.Segalla-Panorama Storico del Nuovo Testamen
to-Queriniana Ed.-Brescia 1984,p.75ss.

110-R.Pacifici – Midrashim -Marietti Ed.-Brescia(Italy)
1986,p.159 .
111-Ekal = Temple or palace, and also one of their
parts : see R.K.Harrison-Biblical Hebrew- NTC Publishing Group - 4255 West Touhy Avenue-Lincolnwood
(Chicago)1993,p.150 .
112-F.G.Nocke-Escatologia-p.126ss .
113- Pubblicazione Settimanale <SPAZIO> - Hobby &
Work Publishing-Milano(Italy)n.23(2013)p.272-273 .
114-P.Grech-Ermeneutica,p.429 .
115-Mc Carthy et al.-Teologia del Patto,p.9 .
116-Mc Carthy,p.10 .
117-P.Grech-p.42 .
118-Mc Carthy,p.28 .
119-P.Grech,p.423ss .
120-G.Colzani- Se Tu Conoscessi il Dono di Dio: see
G.Ravasi et al.-Fede e Cultura,p.87 .
121-G.Colzani,p.98 .
122-Sanna-L'Uomo...,p.397 .
123-G.Ravasi- Principi di Esegesi: Introd. al Vangelo di
s. Giovanni: see Fede e Cultura,p.29 .
124-M.Clevenot - Gli Uomini della Fraternità ,4° Vol. -
I Cristiani al Tempo di Maometto - Borla Ed. - Roma
1984,p.9 .
125-P.Duffet Smith – Practical Astronomy With Your
Calculator-Cambridge University Press-N.Y. 1982,p.5

126-G.Romano – Introduzione all'Astronomia-Franco Muzzio Editore-Padova 1985,p.230 .

127-The World Almanac And Book Of Facts 1997 - #1-New York-Times Bestseller-One International Boulevard Suite-630 Mahawah-New Jersey 07495-0017, p.476 .

128-G.Romano.

129-Pubblicazione Settimanale <SPAZIO> - Hobby & Work Publishing-Milano(Italy)-n.23(2013)p.272-273 .

130H.Robert Mills-Practical Astronomy-Albion Publishing-Chichester 1994,p.110: see the practice (=method of Saros) of the periodic eclypses with a simple diagram in G.Romano .

131-J.Gillings - Mathematic in the Time of Pharaohs-Dover Publications Inc.-New York .

132-A.Parrot-Archeologia della Bibbia Newton Compton Ed.-Roma 1978 .

133-H.R.Mills - Practical Astronomy- A User-friendly Handbook For Sky Watchers - Albion Publishing - Chichester 1994,p.88 .

134-R.North-Geografia Biblica:see Grande Commentario biblico-Queriniana Ed.-Brescia 1974 .

A.Rolla- La Bibbia difronte alle ultime scoperte – Pauline Ed.-Roma 1965 .

135-Dizionario Biblico-Diretto da F.Spadafora-Editrice Studium-Terza Edizione-Roma 1963,p.587.

Printed in April 2019
by Youcanprint